T0182121

Managing More-than-Moore Integration
Technology Development

Riko Radojcic

Managing More-than-Moore Integration Technology Development

A Story of an Advanced Technology Program
in the Semiconductor Industry

With contributions from Matt Nowak
&
Artwork by Adam Radojcic

 Springer

Riko Radojcic
San Diego, CA, USA

ISBN 978-3-030-06495-2   ISBN 978-3-319-92701-5   (eBook)
https://doi.org/10.1007/978-3-319-92701-5

Printed on acid-free paper

This Springer imprint is published by the registered company Springer International Publishing AG
part of Springer Nature.
The registered company address is: Gewerbestrasse 11, 6330 Cham, Switzerland

Acknowledgments

First and foremost, all the contributions, which were so much more than just the constructive feedback and encouragement, by Matt Nowak – my friend and a colleague – must be acknowledged. In addition, the artwork, including the illustrative cartoons and caricatures, used throughout the book for emphases, was created by Adam Radojcic – my son.

This book is very different than anything that the author has done in the past, and the encouragement and support received from friends and family was vital and is greatly appreciated. Special thanks are due to Prof. Andrew Kahng of UCSD, Prof. Shirin Hooshmand of SDSU, Mariano Llosa, Evan Edwards, and Irina Struckus for their excellent feedback. And of course, this book would not be without the encouragement and the constructive inputs from the family – my children who have become such wonderful adults, my brother, and nieces… Thank you, one and all.

Preface

This book is an *engineering management* book. But different. It presents real engineering management challenges and shares real hands-on experiences and insights accrued over a 30+ year career in the semiconductor industry – but using a novelized format. The material is presented through a fictional story that follows a narrator as she grows from an intern to a professional, in a made-up fabless semiconductor chip company. Her story describes the management issues encountered in running a technology development program, as well as the work-life balance challenges experienced by engineers working in the technology industry. It is presented through a series of realistic vignettes which also describe the dynamics and the atmosphere typical of working in the semiconductor industry.

The story is developed in three sections separated in time by a year or so, and corresponding to a period when the narrator is an intern, then a young engineer, and finally an experienced technologist. These time frames also correspond and correlate to an Advanced Technology Integration program as it evolves through the definition, execution, and delivery phases, spanning a total of about 5 years – a minimum period required to track a semiconductor technology development program. In addition, each section also includes a separate chapter that shares the thoughts of the key characters in the story and provides the views on the More-than-Moore Advanced Technology Integration program from a technologist's, a manager's, and a businessman's perspective.

'More-than-Moore' is a name of a technology paradigm that is expected to succeed the practice that the semiconductor industry has followed for the last 50 years – commonly described by the well-known 'Moore's Law'. The Advanced Integration Technology program described in this story is focused on the More-than-Moore type of technologies. However, the experiences and insights described are believed to be applicable to development of any disruptive technology, i.e., any technology that is not directly tied to a mainstream product evolution pattern.

The book is intended for a nonexpert reader. Hence, the technical background that enriches the reading and offers insights in silicon technology and semiconductor industry is segregated into a set of 'Technical Background Boxes' (TBB) – highlighted in gray for a motivated reader. These may be skipped by other readers. In addition, a set of illustrative cartoons, highlighted in green, are included as a light-hearted summary of the serious topics addressed in the story, including the people-, technology-, and corporate-management issues.

All this in order to share real-life engineering management experiences in a light and readable way, with both, the readers from the semiconductor industry interested in the specific insights in managing a More-than-Moore technology development program, and the readers from outside the industry interested in general technology-management practices. Enjoy!

San Diego, CA, USA   Riko Radojcic

Contents

Part I   I, Intern (and Setting Up an Advanced Tech Integration Program)

1   The Interview   3

2   The Start   7

3   The Background (and Beer with Cz)   17

4   The Challenge (*Why* Do Advanced Technology Development?)   33

5   The Value (of an Advanced Technology Program)   39

6   The Knobs (for Setting up an Advanced Technology Effort)   43

7   The Proposals   55

8   The Good-Bye   63

9   a Technologist: Cz's Contemplations (Circa Year 0)   69

Part II   I, Engineer (and Running Integration Tech Development)

10   I'm Back!   83

11   Managing Learning (in a Matrix Organization)   93

12   Managing Compensation (in a Matrix Organization)   103

13   Reorganization (into Line Organization)   107

14   The Challenge (*How* to Do the Integration
       Technology Development)   119

15   Test Chip Engagement (and the Assignment I Did Not Get)   131

16   PathFinding Engagement (and the Assignment That I Did Get)   139

17   a Manager: Steve's Soliloquy (Circa + 3 Years)   145

Part III   I, Leader (Product Intersection and Handing Off the Learning)

18   I'm Back! 155

19   Process Transfer Case Study (IPD) 161

**20   Methodology Transfer Case Study
        (Multi-Physics Stress Modeling)** 167

21   Concept Transfer Case Study (2.5D Integration PathFinding) 181

**22   The Big Picture (Development of More-than-Moore
        Integration Technologies)** 193

23   a Businessman: Mao's Musings (Circa + 5 Years) 205

Part I
I, Intern (and Setting Up an Advanced Tech Integration Program)

This section finds Jasmine – a narrator of this story – joining a fictional semiconductor company as a summer intern. Through her narrative, some of the background history of the evolution of the semiconductor industry, as well as the basic industry terminology and structure, is described. The team that she has been assigned to struggles with defining a program for advanced technology development and integration. Jasmine discovers that the hard part is knowing what you want.

Chapter 1
The Interview

"Hi, my name is Jasmine"

"Come in, sit down Jasmine. Can I get you anything? Water? Coffee?"

I sit and worry. I am a grad student, interviewing for a summer intern position with RoCo Inc. Everyone tells me that internship is a shortcut to a real job offer – once I finish the damn thesis and actually graduate – so I am nervous. For all I know, my first job, my career, my entire life, and my future, all hang on this, my very first step in the semiconductor industry. So, I worry. And RoCo Inc. is the company to be with, at least in the area of chip technology – the topic of my research work at the university. So, I worry. And, it *is* in town. I like San Diego. I would like to stay in San Diego. Victor likes San Diego. I *need* to stay in San Diego. So, I worry. And my prof – aka 'prof Dracula' – tells me that this dude, a Dr. Czochralsky, is the man to impress. So, I worry. And I am a girl applying for a position in boy's club. So, I worry….

"Relax, Jasmine. Tell me about your graduate work."

Easy for him to say. Relax!?! Dr. Czochralsky is an old guy, with a bit of a reputation. I read some of his papers and saw him talk at several conferences. All the gossip is that he is a bit of an oddball character but that he should be listened to – partially because he is with RoCo and partially because he is supposed to be a bit of a visionary. Looks sort of stereotypical – a bit fat, wears a jacket with sleeves that have become shiny with use, thick glasses on a crooked face, gray hair that is in a need of a cut, and a graying cropped beard… His office is also typical. White board dominates one wall, various framed photographs on the wall above his desk, window looking out over the parking lot, and charts and graphs randomly hanging everywhere else. He puts his feet up on the table – showing off well-worn deck shoes – and says again:

"So, what are you working on with the prof? You know, you guys may think him a bit of an OCD slave driver, but he does lead one of the more innovative research groups in our technology."

© Springer International Publishing AG, part of Springer Nature 2019
R. Radojcic, *Managing More-than-Moore Integration Technology Development*,
https://doi.org/10.1007/978-3-319-92701-5_1

"Errr, I have been looking at the effect of the exact shapes of transistors on performance and the impact that this has at circuit level. I have some papers that we published…" I say reaching down in my purse.

"Yes, I think I saw those, and I am sure you know a lot more about that than I do. Let's talk about the basics. Tell me, Jasmine, what is a transistor?"

"A transistor? Well, Dr. Czochralsky..."

"Call me Cz. Everyone does."

"OK, Dr Cz, errr, a transistor is an electronic device, a switch, a 3-terminal device where the current is…".

"Can you draw it on the white board?"

I get up and draw the usual symbol for transistors used in circuit diagrams. He gets up and draws several sketches on the board and asks me if those are also transistors. I explain what they are – just alternative ways of representing different aspects of a transistor – except for the last one that looks to me like a random set of polygons and arcs. I am stumped by that one.

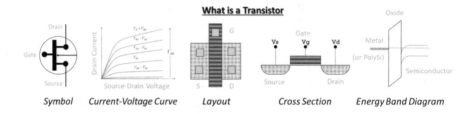

| Symbol | Current-Voltage Curve | Layout | Cross Section | Energy Band Diagram |

"And this one", he asks, pointing to the last sketch.

"Errr, I am not sure about that one. Could you, errr..."

"Good, good. Let's talk about that one then," he says.

Damn! He wants to talk about the one thing that I am confused about. Just my luck! Or is he just playing with me? Maybe because I am a girl? What the hell is it anyways? What class did I sleep through? So, I squirm, and I am sure that by now, I am dripping with sweat. Whoever said that girls only glisten rather than sweat is an idiot.

"Don't worry Jasmine," he says. "I don't want to talk about the stuff you know. I want to see how you do with things you do not know – how do you deal with questions when there is no back of the book with all the answers. After all, the things you know now are going to be obsolete in a few years. In our business you have to be learning new stuff all the time. I want to hire someone for what they can be, not what they already are…".

"Errr," I stammer, not quite sure what to say.

"The fact that you are doing graduate work is great – it shows that you are curious and like to learn, and therefore you are probably not the kind of a person who goes into keyboard-lock mode when there are no ready answers. This is the kind of people we are after," he continues. "In our industry, especially if you are working in a development team, you work on stuff where there are no answers. No one knows. Not I, not prof Dracula, not the engineer next to you, no one… Does this scare you or light you up?"

"Errr… I don't honestly know," I stammer. "So far, there has always been someone who knew, or at least someone who knew more than I did... But I am willing to try" I add, putting up a brave face.

And so it goes. Turns out the sketch of lines and arcs that he drew on the board is an energy band diagram – just a solid-state physics way of representing a transistor. The last class that I took in solid-state was when I was an undergrad, so it just did not come to me. He asks about the electrostatics of a transistor, how they are made, what are the various manufacturing process steps, what are design rules, what are models, and on and on... Basic stuff that I have thought of as background noise and did not really focus on. He asks about the end of Moore's law, packaging, test, etc. He uses terms and names that I am not familiar with. I, of course, question what he means by this and that, and he explains. And so on – it's mostly him talking and me asking questions. It started feeling a bit like I was back in grade school, but after a while, I got into it. There is a bunch of things that I knew little about and that I never thought to question. The way he puts things, it seems like everything is connected, one thing leads to another; a complex web of technical – and often business – trade-offs... Interesting… The business side of things – totally new to me.

In the back of my mind, I am sure that I bombed the interview. Feels like I did not know most of the stuff he asked. I am bummed by that, but instead of fading and tuning it out, like I normally would, I get into the discussion, feeling stimulated and somehow more at ease with this Dr Cz. I actually start wondering about why this, how that, and so on, and we end up having a great conversation. In what feels like a few minutes, but was really a whole hour, he says:

"So, when can you start?"

Chapter 2
The Start

That was a couple of months ago. RoCo Inc. did make me a nice offer of a summer internship. It is not only an excellent check mark on my resume but pays much better than the hostessing that I did last summer and includes crazy perks like surfing lessons, microbrew tours, and stuff like that. Ha! Maybe this education gig really does pay. I accepted. Duh!

Since then I have completed the semester at the U and of course slaved for prof Dracula. He was due to present some of our results at one of what seems like an endless list of conferences that he attends and pushed the team to complete the analyses, do extra work, polish the paper, prepare the slides, and even assist in organizing the conference. This of course meant many hours in the labs, many more in the office, many late nights, and the inevitable e-mails from prof Dracula at 3:00 am – demanding this or that… The man never sleeps, it seems. This plus his life-sucking demands earned him his moniker: prof Dracula. Deservedly so! And, as usual, this workload also meant that I haven't visited mom and dad up in LA for what seems forever, and I had to break a few dates with Victor.

Victor is my boyfriend. He is a grad student in the Life Sciences department. We met a few years ago – when we were both just starting our grad studies – at one of those university student gatherings. And we seem to have hit it off and have been dating ever since. He is a cute all-American kind of a guy, quite intelligent and articulate, and very sweet. The fact that he was brought up as an east coast WASP – unlike me who grew up in LA, a daughter of a Latino car mechanic and a Persian mom – does not seem to be a problem. This despite what my mom says. My mom – a whole story onto herself – seems to believe in the old school rules that appear to dictate that if I really have to date – something that should be avoided – it should be only guys who grew up a block from our house, are someone's cousin, and are preferably Iranian. Anyways, Victor and I have been together for a while. He does seem to have tastes that go with his preppie background and are more expensive than what I am used to. But he likes to wine and dine *me* and is fun and interesting to be with. We have talked about things we want after we graduate, and I am beginning to feel like he, possibly, maybe, perhaps, may be 'the one'. We'll see….

© Springer International Publishing AG, part of Springer Nature 2019
R. Radojcic, *Managing More-than-Moore Integration Technology Development*,
https://doi.org/10.1007/978-3-319-92701-5_2

Anyway, my internship started this last Monday. Dr Cz met me in the lobby, walked me through the offices, and introduced me to the various people who work in his group – none of whose names or job titles I seem to be able to remember now. One is a woman – Panchali something – so at least I will not be the only girl around here. We met for a bit in his office, and he said that in his opinion, all interns should start in a device lab – mentioned something about this being 'where models meet Silicon' – and then walked me to Doug's office. Doug – a Douglas Daicho – is a younger engineer, an Asian, but sounding like the boy next door, who works in the device lab and who will be my direct supervisor. Skinny and somewhat weedy looking man in his mid-30s, I guess. Classic nerd look: dark pants, white baggy shirt, black leather shoes, thick glasses, and a series of electronic devices clipped to his belt – a couple of phones and an old school calculator. No pocket protector, however. Seems very earnest and serious – a bit intense. He stood up and almost saluted – military style – when Dr Cz came into his office. They talked about the various projects that they could start me on and ended up agreeing on something they called 'Titicaca'. Doug then rummaged around his desk, cluttered with file folders, papers, books, and strange bits of hardware, and gave me a fat binder, saying that I should read the documentation before I start in the lab.

He then walked with me to what is to be my office – a cubicle in a maze of cubicles, in a large open floor plan office. It really is like in the movies – like 'Office Space' – a sea of nondescript beige and brown, with a constant background buzz of computer fans, muffled conversations, phone dings, air-conditioning hum, and a high-frequency whine of the overhead lights. I worried whether I would be able to find my way back to my cubicle, if and when I have to go to the bathroom. And of course, as always, I worried whether everyone will see a girl or an engineer.

Good news! After a few days here, I have discovered that I can safely go to the bathroom and find my way back through the maze. I now know where the coffee machine is, I have met some of my cube neighbors, and I am beginning to believe that I will not die – at least not just yet. I met – again – most of the people who work in Dr Cz's group, had lunches and coffees with some of them, attended a few meetings, had a tour of the labs, and generally started getting used to this place. It's all right – so far.

Turns out 'Titicaca' is a name of one of their test chips – a chip that contains discrete electronic elements so that their electrical characteristics can be measured directly – as opposed to an integrated circuit, where all the transistors are interconnected internally to make a memory or a processor or any one of those ICs that we have been studying at the U (*TBB 1.1). Apparently, every time a new generation of technology is brought up, RoCo designs a set of test chips to evaluate the manufacturing technology and to assess the performance of some of the basic circuits it plans to use. Sort of like the car companies use concept cars to evaluate new design ideas or new engines. Titicaca is a name of one of those. Victor – the Life Scientist among us – said over dinner a couple of evenings ago that the way these test chips are used sounds a bit like biologists studying snails' brains in order to understand the workings of neurons in human brains. Simpler creatures seem to allow access to single nerve cells and stuff – same as test chips allow access to single transistors and stuff. Hmm, interesting analogy...

Technical Background Box 1.1: IC Product Types

- *Memory* – an IC (or a block on a chip) intended to store data. There are many different types of memories using different data storage mechanisms and leveraging various technologies. Classification taxonomy includes:

  - Volatile vs. nonvolatile memory, based on ability to retain the data in power-off mode
  - Static vs. dynamic memory, based on ability to hold data in steady state and without refreshing
  - Read only vs. read and write memory, based on the data programing mechanism
  - Flash vs. random access, based on the programmable block size and access
  - Electronic vs. magnetic vs. resistive memory, based on the fundamental storage mechanism
  - Stand-alone vs. embedded, based on the use mode and proximity to processor functions

- *Processor* – an IC (or a block on a chip) that executes various OS and software instructions. Often referred to as the "brain" of the electronic appliances since it is used as the master controller of all sub-systems and manages all the data flow. There are basically two types:

  - Central processing unit (CPU) typically optimized for rapid execution of generic instructions
  - Graphics processor unit (GPU) optimized for processing of large blocks of data in parallel

- *Application-specific IC* (ASIC) – an IC type that is optimized for a specific custom application rather than for general-purpose use. ASIC also sometimes refers to a type of chip design style – such as is used to define this class of ICs. There are several classes of ASICs, but typically they fall in two categories:

  - Gate array – where the placement of logic gates is fixed and the customization is implemented only in the interconnect layers
  - Standard cell – where the customization is achieved by suitable placement and interconnect of pre-designed logic gates, embedded memories, macros, etc.

- *Field programmable gate array* (FPGA) – an IC (or a block on a chip) designed to be configured by a user after manufacturing. There are several types based on the fundamental mechanism used to program a specific configuration (SRAM memory vs. floating gate vs. fuse, etc.)

(continued)

Technical Background Box 1.1 (continued)

- *System-on-chip* (SoC) – an IC that integrates many components of a system and may contain digital (often including multiple processors, such as CPUs, GPU, and specialized processing engines, memories, random logic, etc.), analog, mixed-signal, and often radio-frequency functions, plus the intercommunication fabric, etc. – all on a single chip.
- *Analog* – an IC (or a block on a chip) that processes analog (vs. digital) voltages and/or currents that vary continuously in time (vs. managing 0's and 1's). Analog designs and technologies tend to be specialized for a given application. There are many types ranging from RF, power management (PMIC), power amplifiers (PA), digital-to-analog (DAC) and analog-to-digital (ADC) converters, etc.
- *Microelectromechanical system* (MEMS) – a chip that also includes mechanical components and moving parts such as levers, cogs, membranes, etc. typically used as some kind of a sensor

Note: Material in the gray boxes is intended for those who are interested in more semiconductor technology and/or industry background information and may be skipped by those who are not.

The binder that Doug gave me describes the purpose of the various test structures on Titicaca, their design, their location on the test chip, the intended test methods, and so on… It did not take me long to realize that it would take me about a year, so to speak, to read it all.

"Don't worry, Jasmine, you do not need to know it all. I just want you to be familiar with the overall architecture of Titicaca and the organization of the documentation." Doug is standing in the door of my cube – basically a gap in the partition walls. I wonder how long he was there and whether he is reacting to what I suspect is a perplexed look on my face. I wonder if he will be a 'protector' or a 'write-off' or a 'creep' kind of a man. These are my names for the typical roles that most guys seem to assume when confronted by that rarest of the species – a girl engineer. Protectors assume you need a lot of help – because you are a woman. Write-offs assume that no amount of help is adequate – because you are a woman. And creeps just stare at you – because you are a woman. Or, maybe, hopefully, he will be none of these – and will instead be just a teacher who will help me be as good an engineer as I can be. My 'sensei'. We'll see...

"You will start by looking at these structures," he says pointing me to a specific chapter, "so that is the section you should concentrate on."

He fidgets a bit, looking like he is about to get up and leave, so I say "I have been going through this tome of yours, and decided to make a sort of a dictionary of the terms that come up. Can you help me with some of the definitions, please?"

"Sure, of course. We do use many acronyms and names and forget that they are not universally known. This is a part of our tribal knowledge, so to speak. Most large companies tend to have some versions of this – almost like dialects of a language," he says. "Show me what you've got."

"Errr… before you laugh at me and my ignorance, let me remind you that most of my grad work has been on parametric yields and none on this process technology or industry structure," I say, "so some of these may seem stupid, but …".

"You know what they say, Jasmine – no such thing as stupid questions – only stupid answers. Show me."

So, I give him the two lists that I have compiled over the last few days – one of the industry terms:

Lexicon of General Industry Terms

- _Technology_ (alt. Semiconductor technology) – *a fuzzy generic term referring to a collection of technical skills and knowledge or processes used to make semiconductor chips. Specific meaning seems to be context dependent – as in 'process technology' vs. 'design technology', etc.*
- _Process_ (alt. process technology, technology, manufacturing technology) – *a sequence of manufacturing steps taken to make a Silicon wafer. Encompasses a series of 'process modules' – basically tightly coupled process steps that perform a given operation, which are then integrated to comprise a complete 'process flow'.*
- _Wafer_ (alt. Silicon, Si, hardware, slice) – *a disc of single crystal Silicon that contains the chips and that is taken through the manufacturing process steps. For Titicaca these are 12″ pizza-sized shiny discs, about 1 mm thick, that look a bit like gold-plated mirrors with a subtle grid-like pattern etched on them.*
- _Transistor_ (alt. Device, MOS, MOSFET, FET, etc.) – *basic electronic element in ICs. There are 10s of billions of transistors on each modern integrated circuit (IC).*
- _IC chip_ (alt. die, Si, hardware, device, component, IC, integrated circuit) – *a rectangular section of a Silicon wafer that contains an integrated circuit. There are typically a few hundred of identical die on each wafer. At the end of the manufacturing process, wafers are "diced" along the "scribe lanes" (alt. kerfs, streets – basically space between the die) to end up with the individual Silicon chips – typically somewhere between few mm's and a couple of cm's on a side.*
- _Design_ (alt. design technology, architecture, layout) – *another fuzzy generic term with several overlapping meanings. As a verb it refers to the activity of defining the function and performance of an IC – i.e., what it does, how fast it does it, etc. Can be applied at different levels – e.g., layout, logical, physical design, etc. As a noun the term describes the end result, i.e., a specific structure on a chip or whole chip or type of chip – e.g., low power design, big design, etc.*
- _Device_ (alt. component, transistor, chip, system) – *yet another fuzzy term whose meaning is context dependent. Refers to transistors when talking about a chip, but can mean the whole IC when talking about PCB or system, and could mean the whole system when talking about appliances sold to end consumers…*
- _IP_ (alt. Intellectual Property, block, library, standard cells, macro, etc.) – *building blocks used to create a given design – usually pre-designed and performing a specific logic function. So, a 'design' comprises a bunch of IP blocks. The term IP is also used to describe any proprietary knowledge – more like the usual legal definition.*

- *Fab* (alt. foundry, factory) – the manufacturing facility used to build the Si wafers, including the clean rooms and all the manufacturing equipment. Modern fabs apparently cost ~$10B and produce 30,000 to 50,000 wafers per month – i.e., big and expensive!
- *Package* (alt. BGA, flip chip, etc.) – a kind of a box that contains the IC chip. Package is usually a thin (few mm's), rectangular (few cm's on a side) molded plastic thing, with an array of pins or 'bumps' on one side – which provide electrical access to the chip. The pins are contacted to test the IC and/are eventually soldered to a PCB for connection to other elements in the system.
- *Board* (alt. Printed Circuit Board, PCB, motherboard) – a panel that contains a set of packaged chips and other electronic components and the metal traces that interconnect them.
- *System* (alt. device, PC, phone, etc.) – the end electronic appliance, such as the phone or a PC, that includes the PCB and other hardware elements, such as a hard drive, keyboard, mice, display, cooling fan, camera modules, other sensors, etc.
- *Foundry* (alt. fab) – a company that does just the Si process development and wafer manufacturing but does not do design and does not have any IC products of its own. Basically, provides wafer manufacturing services and gets paid per manufactured wafer.
- *OSAT* (Outsourced Assembly And Test, assembly house, packaging company) – a company that sells services to saw the wafers into die, to assemble die in packages, and to test the packaged ICs. Sort of like a foundry – but for the package-related manufacturing steps rather than Si processing. So, foundry + OSAT = complete chip manufacturing flow.
- *Fabless* Co (alt. design house, Integrated Fabless Manufacturer) – a company that defines, designs, and sells ICs but uses an external foundry and OSAT to manufacture them, i.e., it does the design only and contracts out the manufacturing.
- *IDM* (alt. Integrated Device Manufacturer, merchant semiconductor entity, vertical semiconductor company) – a company that does both the design and manufacturing of ICs, i.e., it has internal Silicon fabs and develops proprietary process technologies used to build their chips.
- *EDA* (alt. Electronic Design Automation, CAD, design tools) – collective name of the companies that make software CAD tools used in the design of chips. Sometimes refers to the software tools that they sell or even the design methodologies used by these tools – as in 'EDA technology'.
- *Moore's law* – an observation, which became a standard practice, that dictates that every couple of years a new generation of process technology is developed to allow integrating roughly twice as many transistors per chip as the existing process technology.
- *Scaling* (alt. shrink) – a practice of reducing the size of features and devices on a die, in order to allow packing more of them on each chip. Thus, Moore's law doubling the number of devices is nominally achieved by scaling their dimensions such that their area is approximately halved.

- *Technology node (alt. process generation, technology generation) – a classification of the manufacturing process technologies and a measure of the complexity and point along Moore's law curve where a given technology fits and roughly the vintage when it was released to manufacturing. Denoted by a number that used to correspond to the minimum feature size in a given generation – like 1um, 180 nm, 65 nm node, etc. Generically, a current generation is called 'n' node, then the next one is 'n + 1', the last one was 'n-1', etc.*

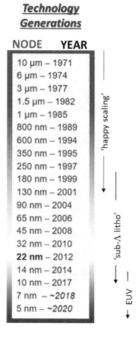

and one of the technical terms (*TBB 1.2).

Technical Background Box 1.2: Lexicon of Technical Terms

- *Test chip (alt. TEG, test slice, etc.) – a chip designed to allow direct access to various elements used to build an IC (discrete devices, interconnect constructs, individual layout features) so that their electrical (and sometimes physical) characteristics can be directly measured. Test chips are not a commercial product and are typically not sold but are used for essential engineering purposes. There are several broad classes of test chips – typically: (a) those used for technology characterization, (b) those used to validate the design IP, and (c) those used as a part of manufacturing process control. The 'characterization test chips' are usually whole die, whose test data provides general technology characteristics and is used to define*

(continued)

Technical Background Box 1.2 (continued)

> the design rules (which dictate the permitted feature dimensions and define
> the disallowed layout constructs) and device models (which describe
> device electrical behavior to be used in design of the actual product ICs).
> The 'IP test chips' are typically also whole die, but they are designed like
> mini-ICs which contain building blocks such as memories, standard cell
> libraries, etc. The 'Wafer Acceptance Test' (WAT) test chips are usually a
> subset of typical structures inserted in the scribe lanes between the product
> ICs and are used to ensure that the manufacturing process is in spec and
> under control.
>
> - *Layout (alt. physical design, polygon design, etc.) – term used to describe
>   the baseline physical design and the drawing of the shapes and features
>   used in an IC. For each of the layers used to build up an IC, it consists of
>   a bunch of polygons, and the overlay looks like this: (different colors and
>   shadings represent different layers)*

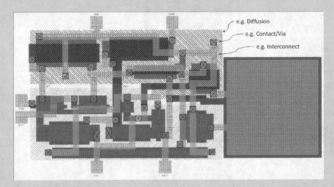

> - *Pads (alt...) – a feature on a chip large enough to contact with a probe for
>   purposes of prepackage test, or to connect to the interconnect used in a
>   package (wire bond or solder ball or Cu pillar, etc.). Pads are typically a
>   rectangular metal feature that is of the order of ~100's of μm on a side, i.e.,
>   huge relative to the rest of the features used in an IC.*
> - *Tape Out (alt. GDS2, etc.) – procedure for releasing the final design data-
>   base to a foundry for mask making and manufacturing – carried over from
>   the time when the design database was really stored (and shipped) on a
>   magnetic tape. GDS is a standard format for describing the various poly-
>   gons and layout constructs contained in the design database.*
> - *Masks (alt. glass, mask works, plates, artwork) – a set of glass plates
>   inscribed with the layout information, which are used to transfer a design
>   onto Silicon wafers using a sequence of photolithography process steps. A
>   complete IC typically requires between 30 and 50 masks to define the lay-
>   ers used to build a chip.*

(continued)

Technical Background Box 1.2 (continued)

- *Lithography (alt. photo, litho, print, etc.) – a set of process steps that use photographic (alt. exposure) and etch steps to define all the features on Silicon wafers. Litho technology constraints dictate the minimum feature size that can be defined on a wafer and is typically one of the bottlenecks to scaling. Litho also dominates the overall wafer manufacturing cost since it is repeated for each mask layer. Other major process modules include depositions (of insulators or conductors), etch (and polish) techniques, and ion implant (or doping) processes.*
- *Yield – a percentage of the die on the wafer that work to spec. Since the cost of manufacturing a wafer is pretty much fixed for a given generation of technology, the cost of an individual IC is a function of (a) the die area (i.e., how many die are there on a wafer) and (b) yield (i.e., how many of those die are actually good). Yield loss is usually categorized in three broad classes: (a) 'functional failures' typically caused by manufacturing defects or contaminants, (b) 'parametric failures' caused by variations in the manufacturing processes, or (c) 'systematic failures' typically caused by interactions between the manufacturing process and specific layout constructs. Normally, for mature technologies and de-bugged designs, functional failures are the dominant loss mechanism - but 100% yield is very rare. Yield is described as a function of 'defect density' – a statistical attribute of a manufacturing process technology and design complexity and die area (big die yield worse than small die because the probability of interacting with a defect - assumed to be randomly distributed – is higher).*

Note: Material in the gray boxes is intended for those who are interested in more semiconductor technology and/or industry background information and may be skipped by those who are not.

"Hmm. Good list of the basics, Jasmine," says Doug. "You know, if you want to know more about the overall industry you should carve out some time with Dr Cz. He is old and a bit of a historian. You may need to humor him for an hour or two for him to tell you all about the industry and technology evolution," he says, with a bit of a smirk. "Maybe a beer with him, sometime. He likes that. As for your Technical Lexicon, I suggest you keep it going because you will be hearing new terms all the time. Sometimes, when I meet with other groups – like packaging or architecture teams – I too am confused because they seem have a whole bunch of terms and acronyms of their own. We all have a lexicon like yours – sometimes in our heads, sometimes as a crib sheet in our notebooks… Any time you are puzzled, just come and ask me or any of the guys. No Prob… Ooops. Gotta run. Meeting…" And off he goes.

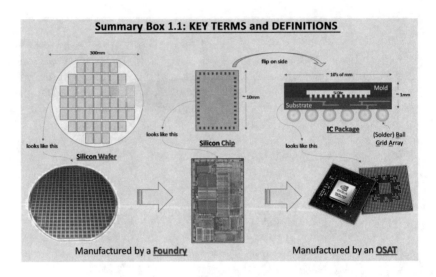

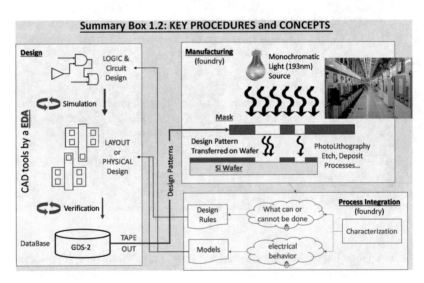

Phew. He did not laugh at me and my Lexicons. Relief! He is all right, this Doug...

Chapter 3
The Background (and Beer with Cz)

"Knock, knock," I say standing in Dr Cz's door – a few days later when I finally found him alone in his office, with the door open – after several attempts of trying to orchestrate a casual bumping into him. Seems like the man is either never there or has someone in his office.

"Hi Jasmine. Come in. How is it going," he says smiling and pointing at a chair while sneaking a glance at his watch.

"Good, thanks. Umm. I have been going through the Titicaca documentation and am compiling a sort of a lexicon of industry terms that are new to me – stuff they do not teach at the university. Doug mentioned that you are good at describing the industry dynamics and the business side of technology. I was wondering if you had the time to tell me more?"

"Of course Jasmine! My favorite topic. Most people are tired of listening to me blabbing on about the semiconductor industry. Tell you what – why don't you join us tonight at Duffy's. I am sure that the guys told you that some of us tend to get together on occasional Friday evenings for a few beers and some gossip," he says.

"Errr. No, no one mentioned anything. Duffy's?" I say, wondering if this is one of those boys-only things where girls are not really welcome – presumably occasions when guys like to beat their hairy chests, grunt and spit and do other man-stuff. And I am wearing my work look – jeans, sweatshirt, flats … and am not dressed for going out. Maybe I shouldn't go? Or perhaps I should?

"Hmm. Maybe they all thought that a pretty girl like you would have better things to do on a Friday night than hanging around with a bunch of nerdy engineers," he says. "Duffy's is a place in the local strip mall just around the corner. Can't miss it. We start trickling in about 6:30. See you there," he grins and dismisses me by turning to his computer.

So, I decide to go – but around 7:00. I don't want to be there first or alone with just one of them – my mom would not approve. Duffy's is a regular kind of a bar/restaurant that is found in any cluster of shops in America. Booths and tables. Dimly lit. Bar along one wall. Couple of TVs hanging behind the bar. Impressive row of beer taps. Random kitschy decoration on the walls – the theme here seems to be

© Springer International Publishing AG, part of Springer Nature 2019

R. Radojcic, *Managing More-than-Moore Integration Technology Development*,

https://doi.org/10.1007/978-3-319-92701-5_3

fishing. Reasonably quiet and empty – at least for now. I am sure it is crowded and noisy when there is a game on or something… I see Dr Cz and a few other guys at the corner table. Dr Cz spots me and waves me over.

"Jasmine come and join us. What are you having?"

"Careful Jasmine," says Doug grinning, "this is a test. If you pick a wrong drink he makes you pay for the round."

"Hmm, looks like you all are on beer. I'll have one of those, too. As long as it is an IPA – preferably local."

Dr Cz waves to the guy behind the bar, holding up one finger, points at his own glass, and says, "Excellent choice Jasmine. If you asked for a Bud-Lite or something like that I would have to fire you. Come, sit down. Let's talk about this glorious industry of ours."

"Thanks," I sit, "this is something that they do not teach us at the university. Fabless, EDA, OSAT, foundries… what the hell are you all talking about? Who are those guys, and why are they?

"Hmm. You want a short version or the whole story," asks Dr Cz.

"Well, I am getting a beer, have no date tonight, and want to learn, so…."

"Excellent. Where to begin?" says Dr Cz scratching his beard. "In the beginning of time – which for our industry is in the 1950's and 60's – it wasn't like it is now. Semiconductor technology was brand new. After all, solid-state transistor was first invented only in late 40's – so the technology was then only a decade or so old. It was a bit of a dark art known only by a select few. The primary drivers were the military and the mainframe guys. So many of the companies playing with the electronic systems in that market – companies like IBM, GE, and AT&T – were looking at the technology in their labs. These, so called vertically integrated companies, were in the business of making complete systems – the computer, all the peripherals, all the components and sub-assemblies, software, hardware, all the associated services… Everything… They thought that the solid-state technology may be an opportunity for making smaller and more reliable systems, and so they had to look at it – at least partially because they worried that their competitors were – leading to a kind of an arms race. So, some even started limited manufacturing of simple solid-state devices – like diodes or single transistors – to replace vacuum tubes in some existing systems. Semiconductors were then these totally strange things, since most of the engineers at the time grew up on vacuum tubes. The technology at the time was just a bunch of empirical cooking recipes. And everybody was doing pretty much everything internally – not only designing and making their own devices, but also developing proprietary process technologies and even building custom process equipment. When I was your age, for example…".

"A thousand years ago?" pipes in Dave – another guy in the group – from across the table. Dave – a David Dupont – is sort of funny. He is in his 50s, I guess, and looks like a regular man-on-a-street of that age: jeans and plaid shirt kind of a guy. Drives a really nice restored Vette, though – noted a daughter of a car mechanic. Wears his hair like it was still the 1970s – except it is gray now and rather thin on top of his head. But he is supposedly a guru in the modeling area – something that is very abstract and mathematical. He just doesn't look like the scientific type.

"Yeah, yeah… laugh on dude – but, if you are lucky, you too will be old one of these days. Oh wait – you already are," says Dr Cz and continues, "no, this was in late 70's. I was a fresh grad – part of the new generation of engineers who came up purely on solid state semiconductors. Anyways, my company had a complete work-shop next to the fab, including a machine shop and even a glassblower, and we home-made much of the equipment and all the jigs and things used in the fab. Furnaces, furnace tubes, wafer boats, and the like. One of my jobs was to help design and build an epi reactor – a giant rf furnace used in the process. Mind you – this was in England – the company was cash-poor and the engineers were cheap by American standards, so it may have been a bit behind relative to the USA. But that was how it was done. All home-made. All internal. The technology itself was obvi-ously much simpler, and the design was done manually – a transistor at a time. And everybody did everything – it wasn't like there were designers and process engi-neers or product engineers, and so on. We were all just semiconductor engineers." He looked almost wistful. Took a swig of his beer and said, "it was actually a lot of fun with all the degrees of freedom to try out all sorts of different things. No restric-tions since no one knew any better."

"You mean as opposed to now, when we are pigeon-holed, and each group is like a separate kingdom. You have to be careful not to step on someone else's toes if you want to do something new?" asks Dave, maybe somewhat sarcastically.

"Yes, but there is a reason why we are the way we are now. Back then – we are talking the 1970's …".

"You were all stoned and practicing free love," teased Doug, and I am thinking that there is a different side to him that is just not allowed to come out during work hours. He is normally so very dry. Must be the Hawaiian part of him – he is Japanese-Hawaiian.

"Ah, the good old days," Dr Cz laughs and continues, "anyways, as the use of solid state devices took off – like I said, mostly driven by military and mainframes – the high priests of the technology went off and started separate semiconductor com-panies of their own. We called them 'merchant semiconductor makers' then, and I guess would call them now Integrated Device Manufacturers (IDMs). Companies like Fairchild or Intel. These companies made just the chips – not the systems. But they did everything for those chips – internal fabs, internal design, internal packag-ing and assembly, and so on. The technologies were still customized and proprie-tary, using internal secret cooking recipes. My bible at the time was a book by Andy Grove – which was truly like a cook book. Very empirical, with diffusion profiles and film thicknesses vs. times and temperatures. Stuff like that – cooking recipes. Andy Grove, by the way was one of the founders of Intel, along with Gordon Moore – as in Moore's law. Sort of funny that these guys who were clearly very technical dudes, started the premier chip company. That was the norm. It was the technologists rather than business or sales guys who started these chip companies. Felt like new companies were cropping up every week in the 1960's and 1970's – especially in the Bay Area. Seemed like, if you knew something about the semicon-ductor technology – rare at the time – got some experience at one of the verticals or existing merchant companies, then, with an ounce of ambition and some funding, you could start a chip company."

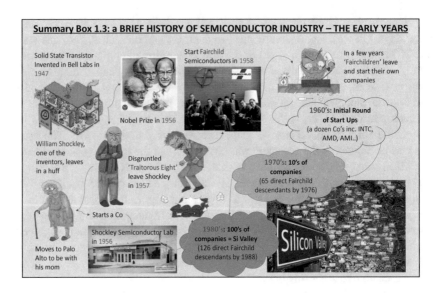

Summary Box 1.3: a BRIEF HISTORY OF SEMICONDUCTOR INDUSTRY – THE EARLY YEARS

"Can't you do a start-up now, too – if you have a good idea and a backer?" I ask.

"Yes you can – but you would have a very hard time getting a few billion dollars required to start a semiconductor manufacturing company now. I guess it is similar pattern in the early days of every transformative technology – you know, a kind of technology that spawns an industry that then changes the world. Like, say, the car industry a century ago. In the early days of internal combustion engine few people had an understanding of the technology, and some of them started car companies in their garage. Ford, Olds, Chevrolet, Opel, Citroen, Porsche… all iconic names and big car brands now – but a 100 years ago they were just a man-in-his-garage kind of shops. In a way it was like that with the semiconductors back in the 60's and 70's. And it wasn't a garage but a clean room. I suppose Silicon Valley is for semiconductors like Detroit was for cars."

"We all up for another round of beers? Maybe some munchies?" Doug cuts in. My glass was almost empty! This is interesting stuff and I did not even notice the beer going. Like folklore stories of old, told by a campfire. "Yes, please," I say draining my glass.

"Yup," says Dr Cz.

"I'll get it," Doug says getting up, "same all around?" We all nod.

"Then," continues Dr Cz, "the pattern described by Moore's law – which was just an empirical observation of what was actually happening in the industry at the time – took off. Partially fueled by a whole range of new consumer products made possible by solid state devices – like cheap and portable transistor radios, TVs, calculators with all of four functions, and stuff like that. Basically, if you wanted to keep-up and survive in the business, you had to double the number of transistors on your chip every couple of years. And obviously – in order to double the integration levels every couple of years without doubling the die size and blowing up your costs, you had to half the size of the structures on the chip. Scaling device sizes was a must. And that necessarily meant that you had to get, or build, new process equipment every couple of years, and that this equipment had to get more complicated every time – and more expensive. The fabs were still internal, though. The mantra was 'real men had fabs'." Said Dr Cz making the air quotes with his fingers.

"Sorry Jasmine, at the time there were very few women in engineering – or in the front offices for that matter. No one ever said real women, or real people, have fabs. It was real *men* who had the fabs. I guess the industry developed its macho culture back then."

"There are precious few women in engineering and front offices now," I say, maybe a bit belligerently.

"True. Sad but true. But it was far more so back then. Lots of women in the industry – but mostly as fab and assembly operators. Moving wafers around the fab, and test and packaging, were all very much manual – this was way before the robotic handlers that are common now. So, all companies had a large population of operators and techs – and many were women. The assembly floor looked almost like cotton mills from the industrial revolution – rows and rows of similar machines, with operators staring down microscopes and manually controlling the process. I remember my managers then musing that only women could do some of those mind-numbing repetitive jobs. But there were very few women in the labs or offices."

"Harrumph," I grunt, "what else is new. Keep the woman down."

"Hmm. Speaking of women in fabs," he remarked, "here is a tidbit that is now totally useless and obsolete. As we all know, cleanliness is essential in wafer processing – hence the fab lines are in clean rooms, and everything is filtered and monitored for highest purity. Cleanliness-technology is an entirely separate discipline in our business – and is foundational for everything we do. Anyways, I remember a number of yield busts back then – that is when none of the chips coming off the manufacturing line work to spec – that were eventually traced back to the operators' make-up. Titanium is used in many make-up products, and flakes of mascara, or whatever, would contaminate the entire production line, causing real panic for product guys and leading to scrapping of the inventory, purging the fab, and all sorts of unmentionably evil things. So, getting lady-operators to give up make-up was a major high-tech challenge at the time," he grinned and shrugged.

"Sounds to me like an urban legend made up just to pick on women," I say, joking. But I feel for those operators of the past. Many women feel like appearing in public without makeup is almost like going out naked. Must have been hard for them.

There is a bit of a lull in the conversation. Then Dr Cz continues – clearly, he is enjoying this. "Anyways, as the complexity of the technology increased, we all had to specialize. It was becoming too complicated for one guy to do it all, so you had to focus on some aspect of the technology and specialize. This was so for the entire companies as much as for individuals.

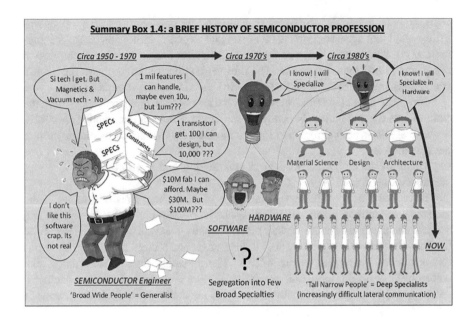

So, the industry basically coalesced into a bunch of specialized layers. Silicon Foundries – specializing in process technology and manufacturing – came up in the 1980s. EDA companies specializing in CAD tools for design, ramped up around then too. This then led to specialized Fabless companies that do design only. Basically, the trend was to shift away from the vertical model where a company does everything to serve a single internal customer and makes, say, mainframe computers, to a supply chain model where a company does one type of a thing only, and services multiple outside customers. The learning curve for those specialized entities was much faster relative to the equivalent activities inside vertical companies – so they quickly got to be not only cheaper but better at what they did. This de-verticalization did not stop at just splitting up chip design from the manufacturing process. It trickled down to specialty process equipment maker companies, mask making companies, IP companies, companies that do fab fixtures and hardware, companies that do materials and gasses, etc. A complete ecosystem of companies, of various sizes, focusing on a given specialty and striving to be the best of breed in a relatively narrow field. Some of these were tiny relative to the old vertical companies. Mainframe makers – like IBM – employed half a million people at their peak. But you could start an IP or an EDA company with few friends and a credit card and be viable with a payroll of 10's of people. Some were larger – 1000's of employees – but in general, big vertical companies were gradually replaced by many smaller, and nimbler, horizontal companies."

"All in the Bay Area?" I ask.

"Well, it started up there. Shockley, one of the original inventors of the solid-state transistor, moved – supposedly in quite a huff – from Bell Labs and the East Coast and started his company in the Bay Area. Why Bay Area? Apparently for no reason other than that he wanted to be close to his parents," he says spreading his hands and shrugging. "This was late 50's. It seems that he was a bit of a crazy tyrant, and his disciples left him and founded Fairchild, which then in turn spawned many other chip companies. The first bunch who left Shockley are referred to as the 'traitorous eight', and many of the second round of startups were led by what were dubbed the 'fairchildren'. You got to love those names – Game of Thrones meets Silicon Valley." He grins and continues, "the local universities – especially Stanford and Berkley – picked up the topic and very quickly became the center of excellence for the technology, further accelerating the agglomeration of the semiconductor industry in the Bay Area. Also, there was already a bit of a start-up culture there – ever since Hewlett and Packard started their company there – before WWII. Hence the famous Silicon Valley. In the beginning the business offices tended to be on the east coast – but after the fairchildren round even the headquarters moved to the Valley. This was in the late 60's and early 70's." He pauses a bit and takes a sip of his beer.

"Off-shoring however started soon afterward," he adds, as an afterthought. "Since a lot of operations were manual and labor intensive, it made sense to move some of the operations to regions with cheap labor. At the time this was mostly to Asia – the Philippines, Taiwan, Korea… Actually – another tidbit of useless trivia – first offshoring, so to speak, was apparently to a Navajo reservation in New Mexico, back in mid 60's. It seems that Navajo women leveraged their rug-weaving skills and became very good at assembly of semiconductor chips… Talk about collision of civilizations…" He shrugs and continues, "But the head office and engineering had to be in the Valley. In the 80's and 90's, and maybe even now, you just were not on the industry radar unless you had a presence in the Bay Area. Back then there were quite a few chip companies in Europe and Japan and the rest of the USA – but they too had at least a business office in Silicon Valley. Everybody was there, and things were really hopping. The saying was that semiconductor engineers in the Bay Area could change jobs without having to change their carpools. Besides, many of the VCs who funded the start-ups were there too. Go where the money is…".

He takes another sip and adds another thought, "Interesting that many of those off-shored entities evolved over time to become independent foundries or OSATs, or whatever, and migrated to the capital-based business model of today, rather than the labor-intensive model from the past. Cost of labor is pretty much a non-issue for manufacturing nowadays. Access to capital and tax benefits seem to be the driver now. And access to engineering talent. Engineering talent is still a big deal – thus ensuring that Silicon Valley remains the center of our business. There is lots of engineering talent there."

The waitress brings a new round of beers and several plates of the usual finger foods – onion rings, carrot sticks and dip, and the like. I take one, take a drink of the beer, and ask, "So, this rapid technology evolution forced specialization and resulted in a break up of vertical companies into a bunch of specialized companies giving us what you now call the distributed supply chain?"

"Yes. Things got complicated and everybody had to specialize. But, if you think about it, this de-verticalization also made sense from the business side of things. Guys-with-ties also liked it. Our industry is pretty much a winner-take-all business.

Early on – in the 70's and 80's – it was the process technology that led to differentiated products. If you had a good process technology then your products were successful, which generated revenues, which could then be used to fund the development of the next generation of technology, which led to another round of successful products, and so on. A virtuous cycle – but uniquely to our industry, a virtuous cycle with a period of only 1 to 2 years. This is incredibly short relative to, say, the car industry, where the development cycle for an engine was 5 to 10 years – maybe longer. This then accelerated the understanding that you had to be #1 or #2 in a given market, or you had to get out of it. With a product cycle of a year or two, the losers were weeded out very quickly. You had to be the best at something to survive. So, you either closed or sold the lines of business that were not dominant in their respective field and focused on what you were good at. So, some of the verticals and even merchants exited the chip business and focused on, say EDA, or spun out Process Equipment Companies, or some other specialty business, or were bought by a bigger rival, or just went out of chip business. And all this was exacerbated by the fact that for every cycle you had to sink an ever-larger capital investment in the fab, since the intrinsic cost of manufacturing was growing. The cost of the process equipment, the fabs, the EDA tools, the masks, etcetera... all necessarily had to balloon to keep pace with the demands of scaled technologies. So, many companies just could not afford internal fabs – they were priced out of that way of doing business."

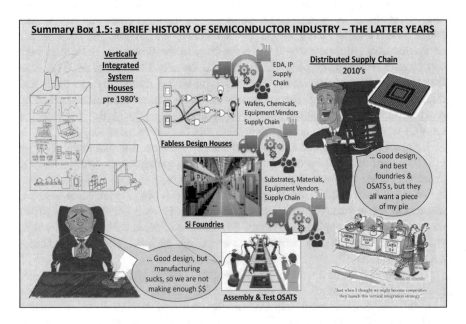

"In a way, this de-verticalization and a distributed supply chain, gave the industry a way of sharing costs across multiple product lines – by spinning out separate third-party entities that serve multiple users. Foundries are a way of sharing the cost of a fab, allowing companies to not have a captive fab. EDA companies are a way of sharing the cost of CAD infrastructure. Etc.… Hence, the industry de-verticalized and asymptoted to a distributed supply chain. The Engineer liked it, The Manager liked it, the VC liked it – and so it happened…".

Another lull. I am wandering if I am getting into an industry that is past its prime. Maybe I should listen to all the talking heads who pontificate at various events sponsored by the Society of Women Engineers, and the like, and go into software? But before I get to ponder more on this, Dr Cz picks up his story.

"Sort of remarkable that all the manufacturing costs went up, but that the price of chips did not increase commensurately," he adds. "That is due not only to the magic of Moore's law, but also to the enormous productivity improvements in the fabs. It wasn't just scaling that resulted in higher value per unit area of Silicon, or the introduction of automation, but also the increases of wafer size – resulting in lower cost per unit area. Since, for a given generation of technology, the cost of manufacturing a wafer is more or less constant, the industry went from 2-inch wafers – when I was your age – to 12-inch wafers now. This was a huge productivity improvement – something like a factor of 50 or so in raw area. I remember reading somewhere that if the car industry went through comparable productivity and efficiency improvements, Moore's law and all, a quarter million-dollar Rolls Royce would cost $250 and go 12,000 miles on a gallon of gas. Or something like that," he mused.

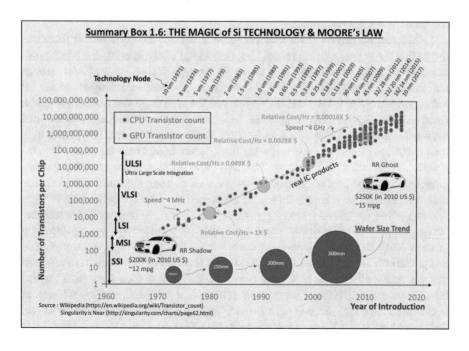

We all pause a bit, sip our beer, munch on what was left of the finger foods, and try to digest all this. "So, this transformation of the entire industry took place in the 80's?" I ask.

"Well, it did not happen overnight. It wasn't an event but a process – and it is still going on. AMD – one of the original fairchildren start-ups – for example, went fabless just in 2010 and spun out a Silicon foundry. TI and Motorola – both iconic companies that pioneered much of the technology – went to the so called Fab-Lite model thenabouts too. Fab-Lite guys keep captive fabs only for specialty products and do mainstream products in foundries. Mind you, fabs for the specialty products tend to be cheaper – because they do not have to keep up with the treadmill of Moore's law. And, of course, the IDMs keep internal fabs. They often dominate their market sectors – like Intel – because they compete in a segment that is large enough to support captive fabs. Intel leveraged the advent of PCs – courtesy of IBM – and focused on making the best CPU for this market, and then evolved to dominate the server market, and so on... Similarly, some vertical companies – like IBM – also continued to maintain proprietary technologies and internal fabs, because they could monetize the differentiation enabled by custom Silicon at the system level, and so pay for an internal fab. They also offset the cost of the fab by offering foundry and other services, to outside third parties. Other companies may have been the best in their respective market segments, but if their segment was not sufficiently rich they could not afford a captive internal fab. So, they went fabless and took the manufacturing to a foundry. It's a dynamic balance between the cost of fabs vs. a given market size and revenues."

"So, the Fab-Lite companies were IDMs but then chose to do some of their product in foundries and some in captive internal fabs? And Fabless companies never had fabs?" I ask for clarification.

"Well, sort of. If you think about it – the advent of the commercial foundries meant that the manufacturing process became a kind of commodity. So, some product classes had to be differentiated primarily through design – rather than through the process technology, as was the case early on. It is sort of analogous to the car industry again. It used to be that the engine was the core thing that made a car model stand out. Not anymore – it is all about the look and feel and the gadgets now – and the engine is a commodity that no one cares about any more. Similarly, some chip manufacturing processes – typically the mainstream digital technologies – became almost a commodity that anyone could buy. The differentiation in those markets therefore has to be primarily through design… Or through attributes like yield and cost, and possibly time to market – after all, some foundries are better than the others, and come up with a technology generation sooner, or sustain faster learning curves. On the other hand, products that are still differentiated by the manufacturing process – like the various analog products – often justify having a custom process and potentially internal captive fabs. IDMs and Fab-Lite guys mostly have products that get differentiation from a process technology, and so they keep their captive fabs. There are still 100 s of legacy and captive fabs out there, making all sorts of components. Not all Silicon is built in the foundries. I think that the recent stats said that something like 50% of chips are built at the foundries, and the rest is in IDMs or Fab-Lite fabs. Memories, analog, CPUs, sensors... all sorts... But everybody uses at least some segments of the supply chain. I don't think anyone is building their own process equipment, or EDA tools, or glassware, etcetera... any more."

"I see. So, companies competing in the digital space focus on design, and companies in analog space still maintain expertise in process?" I ask.

"Well, sort of. That is a bit too broad-brush. Typically, IDMs and Fab-Lite guys tend to have one kind of a corporate DNA, so to speak – with deep process technology roots. Fabless companies tend to have a different corporate DNA because they grew up focusing on just design and using the distributed supply chain. But by now much of these differences have been smeared out and there is a continuous spectrum of business models. Some Fabless companies – typically the big ones – have a lot of technology expertise and may collaborate with the foundries or OSATs to steer the development of process technologies to suit their products. Some companies have proprietary processes that are run at the foundry or OSAT. This is typical for Fab-Lite companies – but some Fabless guys do it too. So, the difference between Fabless vs. Fab-Lite is often not black and white. Unless you are old like me and remember who came from where. The DNA differences show up in how companies relate to risks and make-vs.-buy decisions, and other more subtle tradeoffs."

"I see," I say, still trying to parse it all, "so, there are really two kinds of companies – those that use this distributed supply chain model and those that use the internal fabs?"

"Well, again," Dr Cz responds, "nothing is so simple and straight-forward. So, for example, IDMs may use portions of the supply chain – like the EDA tools, or assembly services from OSATS, either as overflow capacity or because they decided to entirely outsource some things. Everybody does use the distributed supply chain, but not necessarily all of it."

"So, isn't it all now a bit of a chaos with no unifying norms? Like lawless wild west. With this distributed supply chain?" I ask. "How do you coordinate all these separate technologies and processes provided by separate companies. Don't you lose all the efficiencies of specialization through the inefficiencies of coordinating many different suppliers? My dad – a garage manager – spends a lot of his time just looking for the right kind of parts for the cars he works on... And that has got to be far simpler than making ICs?"

"Aaaah. I love the way you think," says Dr Cz and raises his glass. "Yes, this is absolutely an issue. But the industry came together and created a number of bodies whose function is to develop various standards that enable interoperability. Actually, our industry is very remarkable when it comes to collaboration among competitors. The technology is very complicated and probably evolves much too fast for the government regulators and legislators to keep up. So unlike, say in the car industry, our industry is mostly based on self-generated rules and standards. The industry even evolved an International Roadmap for Semiconductors – ITRS – which basically served to coordinate all major development across the entire industry (*TBB 1.3). Serving like a drummer that keeps a rock band in rhythm. It was a magical thing that dictated what needs to happen when, at various levels of technology, in order to sustain Moore's law cadence. It is now pretty much out of gas, since the technology is butting against the end of Moore's law, but back in the 90's and

2000's, it defined not only the target pitches and feature sizes as well as the introduction date for each technology node, but also the areas that require development, and various metrics that need to be met, and so on."

Technical Background Box 1.3: Moore's Law and ITRS

- *Moore's law* refers to an empirical observation that the number of transistors in a (digital) integrated circuit doubles approximately every 2 years. Moore's law is an observation or projection, named after Gordon Moore, and not a physical or natural law. It has been embraced by the semiconductor industry as a requirement for technology development and evolution and has been embedded as the fundamental principle for the National Technology Roadmap for Semiconductors (NTRS) and International Technology Roadmap for Semiconductors (ITRS) (https://en.wikipedia. org/wiki/Moore%27s_law).
- International Technology Roadmap for Semiconductors (*ITRS*) is a set of documents produced by a group of semiconductor industry experts (voluntary), representing the consensus and the best opinion on the directions of research and timelines for the semiconductor technology, with a horizon of about 15 years into the future. The foundational principle was a continuation of Moore's law, i.e., definition of the developments necessary to extend Moore's law. The initial issue (1996) was focused on the USA only. Subsequent issues involved international participation, resulting in the ITRS reports serving as a basic specification for the global industry that was issued every 2 years, or so, starting from 2001. The roadmaps addressed all key aspects of the semiconductor technology, including the following specialties (http://www.itrs2.net):

System drivers/design	Factory integration
Test and test equipment	Assembly and packaging
Front-end processes	Environment, safety, and health
Process integration, devices, and structures	Yield enhancement
Radio frequency and analog/mixed signal	Metrology
Microelectromechanical systems (MEMS)	Modeling and simulation
Photolithography	Emerging research devices
IC interconnects	Emerging research material

Note: Material in the gray boxes is intended for those who are interested in more semiconductor technology and/or industry background information and may be skipped by those who are not.

"And it was organic – in the sense that no one from outside dictated that this needed to be done. The industry came to a consensus and used various industry bodies to develop standards and things like the roadmaps, and so on. Nowadays, with just the engineering effort to develop a new generation of technology costing a billion or two, there are the consortiums, where companies collaborate and share the cost of even the pre-competitive technology development. To IDMs that is like sharing the family jewels. Participating companies donate their engineers, and pay membership to these bodies, to develop basic technologies, process modules or materials. Other collaborative bodies coordinate standards or roadmaps or what have you. I think that this is one of the benefits of the fact that the industry is fairly small and concentrated in few places on the planet – like the Valley – so everybody knows everybody else. So, this kind of things can be coordinated. Quite magical and, I think, unique… But, in principle, you are right, and understanding all the links in the supply chain is a specialty skill of its own," he concludes.

"Geeze. It's complicated and not easy to sort and categorize," I say, giving up trying to make a simple mental model of the industry.

"Oh yes. It *is* complicated. And just to make it more interesting, the industry – or portions of the industry – seems to be coming around a full circle, and there is now a trend towards re-integration to make new vertical or virtually-vertical entities. Some of this comes through consolidation – big companies buying smaller entities, either to enhance their position in a given market or to buy their way into a new market. With maturing technology and saturation of given market segments – like say the PCs – this is a way of maintaining the growth rate that Wall Street has come to expect from companies in this industry sector. So big guys gobble up littler guys.

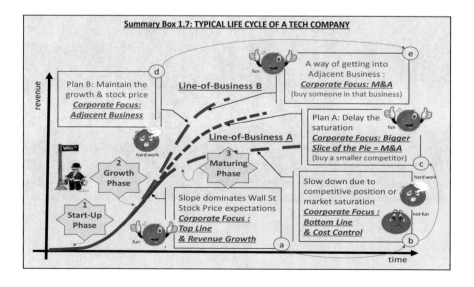

"With the cost of fabs and technology development going into many billions of dollars, this type of consolidation is probable, or even inevitable, for the semiconductor sector – and would have been unthinkable back in the 80's and 90's. Back then, the birth rate of new companies far exceeded the consolidation rate. Now it is the other way around … well, except for entirely new product or technology segments – like IoT or MEMS. There are economic models that say that as an industry sector matures, a balance between government regulations and economies of scale dictates that it consolidates down to three dominant players. Just like the car industry in the USA – with GM, Ford, and Chrysler," says Dr Cz, while counting 1-2-3 on his fingers.

Technical Background Box 1.4: Types of Entities in Supply Chain and Example Companies

Category	Example	Example
System co (current)	Apple (AAPL)	Huawei Technologies
Vertical companies (old)	American Telephone and Telegraph (AT&T)	International Business Machines (IBM)
Internal labs	Bell labs (Murray Hill)	Thomas Watson labs
Conglomerates	Samsung	Toshiba
Integrated device manufacturers (IDM)	Intel (INTC)	Fairchild (now ON semi)
Fab-lite	Texas Instruments	NXP (née Philips + Freescale (née Motorola))
Fabless companies	Qualcomm (QCOM)	Nvidia (NVDA)
Foundries	Taiwan Semiconductor Manufacturing Co (tsmc)	GlobalFoundries
OSAT	Amkor Technology (AMKR)	Advanced Semiconductor Engineering, Inc. (ASX)
Electronic Design Automation (EDA)	Cadence Design Systems (CDN)	Synopsys (SNPS)
Equipment manufacturers	Applied Materials (AMAT)	ASM Lithography (ASML) (JDP by ASMI + Philips)
Industry standard bodies	Joint Electron Device Engineering Council (JEDEC)	Institute of Electrical and Electronics Engineers Standards Association (IEEE-SA)

Note: Material in the gray boxes is intended for those who are interested in more semiconductor technology and/or industry background information and may be skipped by those who are not.

"In addition, there are companies that are re-verticalizing in a given sector as they optimize the integration of hardware and software, and what have you. They tend to do more and more customization to optimize the system – and end up as a virtual vertical entity. Like Apple, for example. At one point they bought everything from the supply chain and focused on system and software design only. But over time they reached deeper into chip technology and now they custom-design their processor chips to mesh well with their operating system, resulting in very elegant integrated solutions. But they still use the foundries and OSATS to manufacture those custom chips.

There are also conglomerates – like Samsung – who compete as IDMs and sell, for example, memory chips, offer foundry services to Fabless companies, sell sub-components like displays, but also manufacture end devices like TVs or phones, or what have you. They do all sorts, and oftentimes are both suppliers and customers in the supply chain" (*TBB 1.4).

"All very interesting, Dr Cz, but I got to go," says Doug, getting up. "Kids soccer tomorrow."

"Yeah, me too," says Dave. They look at the bill, mumble a bit, put some cash on the table, and amble out.

"I guess I should go too," I say and dig around my backpack for my wallet.

"Don't worry about it, Jasmine – this is your welcome aboard treat. But, before you go I have a question for you," says Dr Cz, nodding bye to the others, collecting the cash, and waving his credit card at the waitress.

"Given all that we talked about, what do you think is the key challenge for the future? What do you think a young engineer like you needs to do to boost her career in this industry?"

"Hmm... focus on one thing and be the best at it ... and change sex?" I ask.

"Well, you could do that – not the sex change part – and I am sure you would do well. As long as the one thing you focused on is the right thing, and there is an employer focusing on that same thing in the area where you want to live. But if you think about it, the industry now consists of this distributed supply chain with all these specialized companies excelling at what they do. The big challenge – as you so wisely pointed out – and hence the real opportunity, is knowing how to stitch it all together to make the best product. That is where the gaps can be. So, I think the opportunity is Integration. Think Integration, grasshopper," Dr Cz smiles, looks at the bill that the waitress brought, and gives her his card.

He then continues, "as for the sex change part – it is true that there is discrimination going on. But you know, in my lifetime the industry has evolved from white men-in-suits to nerds-in-hoodies. In fact, Shockley, a Nobel laureate and a founding father, espoused some crazy racist ideas in his later days and the industry that he started whole-heartedly rejected him and his views. Kudos to the industry. Throughout my professional life I felt like it was a bit of a United Nations – and you could meet people from all over. I guess, the industry basically sucked talent from wherever it could find it. Led to the so-called brain drain from many places – a lot of it *to* the Silicon Valley. So, who knows, the next cycle may bring true egalitarianism which includes women too. No one will know or care about the race or gender of whoever comes up with the next killer app, or the next hot chip. So, no need to change your sex. Besides, you are a pretty girl and it would be a waste," he says, smiles, signs the slip, and gets up.

We walk out together. "Thanks, Dr Cz. This was fascinating and very educational. I really appreciate it," I say, ignoring his last somewhat edgy comment. He probably meant it as a complement. Don't make a fuss...

"No worries, Jasmine. I enjoy telling old war stories. By the way, you OK to drive?"

"Pffft. Two beers in three hours. No Prob. Good night. See you on Monday."

And as I drive home, I am trying to control my exuberance and am thinking to myself. "Wow!! This semiconductor industry is like a big complicated machine with many moving cogs and wheels and things. It produces these mind-blowing chips with billions of devices that are a few atoms on a side – amazing and fantastical. But it all seems to work. And it is fueling what my Philosophy Prof – I audited a 'philosophy and science' class last year – called a Second Renaissance. The chip technology has changed every aspect of life – the way we communicate, the way we bank, the way we work, the way we date, almost the way we think. Everything. I do want to be a part of it. Privilege to be in the very center of the engine that is changing the world. And I – a girl – am going to contribute. Yaaas! Little ol' me – the first member of the family ever to get a grad degree – if I ever get it, and a girl to boot, will help change the world. Yaaas!!! I like what Dr Cz said – about from men-in-suits to nerds-in-hoodies and then, maybe, possibly, to someone like me! Ha!"

Chapter 4
The Challenge (*Why* Do Advanced Technology Development?)

So, about a month or so goes by. I have got a hang of working in the lab and getting the data. I now know how to work a prober – basically a setup where a centipede-like fixture is aligned and contacts the pads on the test chip so that appropriate measurements can be made on the corresponding test structures. I know how to program all the instruments to conduct the required measurements. And these measurements virtually involve counting individual electrons and measuring the speed of light. Wow! And I know how to download the data into the lab database and to do the initial sorting and data manipulating. And I am beginning to learn how to create models that describe the device characteristics – which are then used in design. Rad!

I am getting comfortable with the people in the group and actually enjoy them. Doug is very good at explaining how and why various things are done in the lab. Very professional and structured. He has set up a regular meeting with me which is very helpful. I started working with Dave on the modeling stuff. He is truly an old-school nerd, with a sense of humor that is bizarre and hard to get. I think I make him a bit nervous – being of female 'persuasion,' to borrow his phrase. He seems to think that this is very funny. But everyone says that he is a world-class modeling guy who knows his stuff very well, and he is never too busy to explain it to me.

And I normally have my afternoon coffee with Panchali – my fellow woman engineer. Well, she usually takes tea. Panchali is quiet and doesn't say much in meetings but talks a lot when it is just the two of us, about her kids, her friends, her life, the way things were in India where she grew up, and our work – everything. She keeps telling me that I am very lucky to be here. She says that RoCo – and especially this group – is a bit of a country club and that in other companies, and in other groups, things are much more aggressive and competitive. She has apparently worked in several companies with much more of a dog-eat-dog kind of atmosphere, where people look at co-workers as competitors rather than colleagues, and certainly not as friends. Seems like in some other groups at RoCo – especially the ones that have direct responsibilities for products – the pressure is much higher, and people are often too busy, or too competitive, to share information or to spend time bringing up an intern. She says that interns are often treated like a source of cheap labor and are usually given mindless

© Springer International Publishing AG, part of Springer Nature 2019

R. Radojcic, *Managing More-than-Moore Integration Technology Development*,

https://doi.org/10.1007/978-3-319-92701-5_4

house-cleaning types of chores to complete. Well – lucky me! I suspect that she is developing a kind of mother-hen protective feeling toward me and keeps giving me tips on the 'right' way of dealing with people, what to look out for in men, and all sorts of pearls of wisdom, along with feeding me cookies or various Indian sweets that she brings from home. My faves are Gulaab Jamuns – sort of viciously sweet donut holes made from cheese. She is very nice, but we are not likely to become bosom buddies since she is mostly a 9:00 to 5:00 person – due to kids and home stuff.

Altogether – it's all good.

Then, a couple of days ago, Doug comes by, looking a bit flustered, and says "Hi Jasmine. There is an all-hands meeting that Steve called. I don't think you are on the e-mail list of regular employees, but you might as well join us. It's this afternoon at 3:00 pm in Oak Tree conference room."

"OK… Thanks" I say, "what's it about?"

"Don't know. But Steve's all-hands meetings are usually not good news," he says and scurries off.

Maybe that is why he is flustered? But, I am told that bad news first come in the rumor mill, and no one mentioned anything at lunch or in the coffee room. So, at 3:00, I join Doug and the others as we pile in the conference room. There are maybe 30 or 40 people there – some of whom I have seen in the corridors but have not met yet. And then this Steve stands up, and says "let's begin."

Steve – Dr. Stephen Crawley – is Dr Cz's boss. I met him a few times before – first at my interview and then in a few meetings. Looks like a regular manager sort of a guy. Middle aged – I am guessing 45 or 50. Clean-shaven, blond but graying at his temples. Medium height and build – as they say on all the TV cop shows. Open collar shirt, pressed and tucked into his chinos. Neat and tidy. Pretty fit looking. I bet he works out in the gym every morning and plays golf on the weekends. A regular Anglo. Seems nice but also quite busy – as I mostly see him in corridors rushing to some meeting or other.

"Don't worry guys" says Steve "It is not bad news or anything like that. No one is getting laid off or fired, and the bonuses are not cancelled as far as I know. In fact – I think I have very good news for us."

There is almost a palpable sigh of relief in the room, and everyone perks up.

"I met with Behrooz today" he says.

"Behrooz is the CEO" whispers Doug to me, guessing – correctly – that I did not know the name.

"He said," continues Steve, "that our products, like Odie and Otto, have been doing really well, with many design wins, and that we are now competing head on with TekMOS".

"Odie and Otto are code names for the latest generation of our processor chips, and TekMOS is a company that dominates the premium tier – an IDM" whispers Doug, again guessing that I may be a bit lost.

"And that," Steve goes on, "in order to get to the next level, we need to be more aggressive in the technology space. Basically, he said that now that we are competing with the big boys, Wall Street analysts are worrying that we cannot keep up with them on technology. TekMOS is an IDM that relies on its advanced technologies and captive fabs, and apparently there is a school of thought out there that believes that we just cannot catch up with them at the top end of the market – because we use plain-vanilla foundry technologies, which are considered to be a generation, or so, behind the IDMs. And this is now depressing our evaluations on the Street. So, Behrooz needs us to ramp up an advanced technology development program that is a credible – and seen to be credible – alternative to the IDM technologies. He mentioned that there is a parallel effort at the board level that is looking at buying or building a fab and leveling the Silicon process technology playing field vs. TekMOS and other IDMs. But he – Behrooz – thought that this is a long shot and is looking to us for alternative proposals. Something short of going all the way to having a fab." He pauses a bit and then adds, "so basically the question for us is how can we beat TekMOS at the technology game, and make products that are earlier, better and cheaper than theirs, even at the top end of the market, while still being Fabless and relying on the foundries and the supply chain?"

Everybody goes quiet. I gather that this is new and something that has not been talked about before.

"That is a big question" says Dr Cz, "what do you need and when?"

"Good point, thanks. I called this meeting just to get you all thinking. I will schedule a series of brainstorming sessions over the next couple of weeks so that we, as a team, can develop some ideas. Behrooz said he wanted outlines of proposals in about a month. He has asked several engineering groups for inputs, so I presume that he wants to review all possibilities and then downselect a few strategies. So, we need the basics in about 2 to 3 weeks, giving me time to make a presentation, polish it, insert some budgetary estimates and other such management stuff. So, guys – think about it. This is an awesome opportunity for us and I think we should grab it with both hands. Could be a lot of fun and certainly would be a lot of work and a boat load of new opportunities… Any questions?"

He looks around the room.

"We are just not built to be an IDM. I would think that it would take us a decade to grow an internal fab. We could go and buy someone – but digesting an IDM and melding their technologies with our products, and their culture with ours, would also probably take a decade" Dave comments, thinking aloud.

"Agreed. Hold that thought and let's discuss in the brainstorming sessions. Think about it, talk about it amongst yourselves, and let's kick it around and see what we come up with. As far as I am concerned, this is our top priority for next couple of weeks. Thanks guys" says Steve, nods and walks out.

The room fills with buzz as people start mumbling in smaller groups. I look at Doug and Dave, who are scratching their heads, and bounce back to the lab, thinking that all this is beyond me.

That weekend I met with Victor and we went out for dinner. I got all done up – dress, makeup, and nice heels, as opposed to my usual jeans-and-tennies look. I even had my hair done – I like to keep it short and easy, but it is always nice to have it touched up. It felt good to be a 'proper woman', without having to worry if my dress is too tight or bright or revealing or whatever. I must say, one nice thing about Victor is that he takes me out to nice places where I can dress up – makes me feel good to be me. Anyways, he said he has some exciting news and that he wants to celebrate – but wouldn't tell me what it is. We went to Gino's – a small Italian restaurant in Little Italy that Victor has been very fond of lately.

"So? What's the news? C'mon, spill" I say; after we finish ordering, the wine is opened, sniffed, and sipped, and we settle in.

"Ha! Yesterday we had a visitor to our labs. A Dr. Argon Wou! You got to admire the name – I wish I was an Argon rather than plain old Victor. He is an alumnus of the department and a CEO of PetGen. PetGen is the company that I have been telling you about – the one that is using similar protocols as ours to develop this new series of drugs for dogs and cats. Remember? Anyways…" he continues without a pause, "we were talking in the lab, I showed him what I was working on, and he offered me a job right there on the spot. Out of the blue. I was stunned – very pleasantly so – but really shook. I must have stood there for a whole minute with my jaw hanging, like an idiot, before I gathered myself and of course accepted. We shook on it. How awesome is that? This will allow me to carry on my research, do the thesis, *and* get paid. Now, is that grand or what?" He stops to catch his breath, ear-to-ear grin on his face.

"That *is* awesome Victor. Like, really fantastic. Congrats" I say and raise a glass for a clink. "So, when and how does this work? Surely you are going to stay here till you get your degree?"

"Yes yes, of course I want my PhD. But the beauty of this is that I can finish the lab work at PetGen and use some of the results for my thesis. They have the latest and greatest equipment so my data will be that much better. He said that they encourage this collaborative approach with the University, and that they would allow me the time to write up the thesis, and do the defense, when it comes to that. How cool is that, hey?" He is clearly excited. "I am moving up to Davis next month! It's only an hour flight from here – so we can meet weekends and things, and then you can move up when you finish" he says.

And there it is. He seems to have decided that this is the way it will be – without even asking for my opinion or considering my wishes and constraints. We always talked about staying in SoCal, but clearly all that is forgotten now that he has this sweet deal.

"Hmm" I say, maybe a bit coolly, "that is great Victor. I am very happy for you. But we need to talk about moving up there. I am a SoCal girl. My family is here. My life and friends are all here. And the things going on at RoCo are really very interesting and promising for *my* career. I am not sure I want to move."

"Yes yes. We shall talk about it, of course" he says, obviously still too excited to hear what I am saying.

The food comes, and I decide to drop it for now. I don't want to be a buzz-kill. But I also am not sure I am ready to move. I like San Diego. I like being able to drive up to see the family. I like getting together with friends from school and doing the holidays with mom and dad and all of my cousins. San Diego is a couple of hours to LA – and all of my past. Close enough, but not too close so that mom and dad can drop by every week. This *is* my home. And the things that Steve talked about – RoCo moving into technology development – is a perfect opportunity for me. It just does not go with moving up to stinkin' Davis. And the plan that Victor and I talked about a million times was to hang *here* – not move. And Victor deciding without talking to me really stinks. But still. We'll see. Lots of things still need to fall in place. I am a bit pissed at Victor, though. He is just assuming that I am going to follow him.

Chapter 5
The Value (of an Advanced Technology Program)

The following week, there is an e-mail calling for a brainstorming meeting in Willow Conference Room on Tuesday at 1:00. Cool! I am sure I don't have much to contribute, but it is exciting to be invited and be a part of the process. I am looking forward to it.

"I decided to do the brainstorming in smaller groups" says Steve once we have all gathered in the conference room. There are ten of us, including Dr Cz and me. "You just cannot do reasonable brainstorming with 50 people, so I am doing this separately with each of the groups and will compile the results"

"You guys are process-design interface and work on putting together the design infrastructure for new technologies. So, I suppose your ideas for technology development activities would revolve around those topics?" Steve continues to kick-start the discussion.

"We could jump right into the concrete projects, but it seems to me that we should first talk about the overall objectives. This whole thing sounds like a pretty big program, rather than a focused project. What are we trying to get out of Advanced Technology Development?" asks Dr Cz.

"Good point Cz. Well, I suppose we do Technology Development in order to end up with better products" responds Steve. "This could be better products in terms of time to market, or product characteristics – like functionality or power or form factor – or product cost and margin...," and he writes on the whiteboard 'Better Products'.

Then he adds "I just read an article in EE Times – really an interview of the head of Technology Development group at Nokia – where he talked about his goal being what he calls Fast Failing. I presume he means that the goal for advanced development is to discover various technology pitfalls early, so as to either fix them or steer the real product away from them." And he writes 'Fast Failing' on the whiteboard and adds: "so there are different ways that a product can benefit from an advanced technology effort. Think big and broad"

"True. Better product has to be the ultimate goal, but there are variations on the theme, so to speak" says Dr Cz, "for example, we could do Technology Development

© Springer International Publishing AG, part of Springer Nature 2019
R. Radojcic, *Managing More-than-Moore Integration Technology Development*,
https://doi.org/10.1007/978-3-319-92701-5_5

just in order to develop a patent portfolio in the technology domain, and then decide subsequently if and how we would use a given patent in a given product. Sort of an open-ended effort only loosely tied to an actual product..."

"That is something to consider" says Dave. "We are a litigious industry and having a good patent portfolio could be extremely valuable – to fend off companies trying to block us from some technology options, or trolls trying to extract some license fees, or whatever."

"Or to do the opposite – and use our intellectual property to keep our competitors from riding our coat tails, or to charge license fees from others" adds Dr Cz.

"Excellent. So, we could do technology development to develop an IP portfolio, either for offensive or defensive purposes" says Steve, summarizing the discussion. And he writes on the whiteboard 'IP Portfolio (defensive, offensive).'

"Hmm," pipes in Panchali. She is normally quiet and does not say much in meetings, so everyone turns to her. "We could also do technology development just to stay abreast of the industry. To be watching and learning what other companies are doing and to be ready for some new technology twist, if and when that comes up?"

Seems like when Panchali gets excited, her Indian accent becomes more pronounced, with longer, purer vowels and accentuated consonants, and I think that even her head starts bobbing and rocking in a stereotypical way.

"Good point Panchali" adds Dr Cz, now getting a bit keyed up. "This could enable us to be the so-called Fast-Followers – someone who is ready to quickly jump on a new technology bandwagon, without necessarily being the first. That also requires technology development work. You got to watch the industry, pick a few promising candidate technology options, play with them and go up the learning curve, so that a very rapid transfer into product could be done, when required."

"True" says Steve, "this kind of knowledge could also be very useful if the company decides to buy some start-up with IP that is judged to be important in the future. That is one of common ways of implementing a fast-follower strategy." And he writes on the board 'Monitor the Industry (Fast Follower, M&A)'.

"Yes, Doug" says Steve when Doug raises his hand. Doug seems to be a stickler for formal rules of brainstorming and waits to be acknowledged before speaking out. This must be the Japanese in him, I am thinking.

"What about doing technology development to disrupt a market, not to track it? To come up with something wild and wacky and very different" asks Doug.

"Thank you, Doug. Good point" Dr Cz cuts in. "That is a different than focusing on making an existing product line better. There is a difference between wanting to disrupt a market – which is something you do when you are a new kid on the block and just getting into some space – vs. wanting to dominate a market – which is something you do when you already have a market share. Technology Development for the first strategy typically seeks to come up with some revolutionary option, whereas the second strategy requires more of an evolutionary technology."

"All good points" responds Steve, "so, in order to maximize the value to the company, advanced technology development effort needs to be anchored to a product family, or at least to a marketing strategy. And if a company already owns a given market, it is unlikely to want to rock the boat with some revolutionary technology – unless our competitor makes us. On the other hand, if it is trying to get into a market that our competitors dominate, it sure would want to rock that proverbial boat, and hope to dislodge the leader" and then adds on the whiteboard 'disruptive vs. dominant'.

And so it goes. Back and forth. Adding. Elaborating. After a while Steve says "guys, this is outstanding. Thank you. How about if we construct a table to summarize and contrast the points raised," and he goes again to the whiteboard.

"Another guy who thinks the way god and Microsoft PowerPoint intend us to" jokes Dr Cz.

Steve ignores this and starts writing on the board. We all watch. "If you think about it" Steve says "all these really describe the potential value that a company could reap from an advanced technology effort, rather than being the objectives. And reaping these potential benefits would require a different level of investment – in terms of people, capital and expenses – and involve different risks, in terms of probability of success." And as he goes along, he crosses and adds things to the table that he is constructing on the whiteboard, ending up with:

'WHY DO ADVANCED TECHNOLOGY ~~DEVELOPMENT~~'

~~Goal~~ Value	Mode 1: defensive	Mode 2: offensive
~~Better~~ *Differentiated product*	*Dominate existing market* • Evolutionary tech. dev. • Med risk • Med investment	*Disrupt new market* • Revolutionary tech. dev. • High Risk • High investment
IP portfolio	*Prevent lock-out* • Incremental tech. dev. • Low risk • Med investment	*Own some turf* • Aggressive tech. dev. • Med risk • High investment
~~Monitor~~ *Keep up with the industry*	*Fast follower (internal TD, M&A)* • Evolutionary tech. dev. • Low risk • Low investment	*Fast failing (be the leader)* • Aggressive tech. dev. • Med risk • High investment

"There. What do you guys think? Does this capture what we talked about?"

Everybody admires the table on the board and nods.

"You know" Dr Cz adds, a bit hesitantly, "looking at this table, it seems that the one thing that is missing from it is the liability of *not* doing Advanced Technology Development. I mean – what is the cost and risk to the company if we do *not* do some of this advanced engineering?"

"Hmm. Good point – and true" responds Steve. "Although, that side of the argument is always very difficult to present and quantify. It is like with the Health Insurance. We mostly choose health insurance policies based on the cost of the premiums, rather than trying to estimate the probability of getting run over by a bus, or whatever. If we were focused on what you call the liability side of the equation, we would all carry much more insurance. But we don't. That liability is the ultimate motivator, but it is difficult to quantify objectively, and is in the eye of the beholder, so to speak. One person could say that the liability of not doing advanced technology is the value of the entire company, since under worst case scenario a disruptive technology invented by our competitor could surprise us and put us out of business. Another person would say that the liability is actually quite small since coming up with disruptive technologies takes time, and given that there are no secrets in our industry we would see it coming and would react accordingly. And so on. So, estimating that liability is a bit like a religious argument. I would like to stay away from that for now. Although – the point is valid, and we should acknowledge it. Maybe qualitatively?"

And with that, Steve takes his phone out, snaps a photograph of the whiteboard and says "Guys, this was great. Enough for now. Look for another one of these next week"

I go back to my cube thinking… I should be digesting the various pros and cons of advanced technology proposition that we just talked about, and try to capture the essence in my notes, so that I do not forget. But instead, I am thinking more about me and Victor. I have been in a bit of reflective mood ever since Victor said he was moving up north. Maybe I should think through what it is that I want – not what Victor wants. What is best for me – rather than for us. Maybe I should consider concepts like investments and risks & benefits – sort of like we just did in the meeting. Investments – of emotions or time or effort? Benefits – like successful career or happy family life? Something… But it is not as simple as Victor vs. Career. After all, I could have some kind of a career up there too. Or I could focus on continuing my time at RoCo, and maybe break up with Victor and potentially find Mr. Right here. But each path has different emotional and physical investments and different risks or benefits…

"What is it that *you* want, girl, and what are you willing to do to get it" I am thinking to myself – and hopefully not saying it aloud, as I sometimes do when in this kind of mood. Those are the primary questions. As trivial as that may sound, it is not easy, really…

Chapter 6
The Knobs (for Setting up an Advanced Technology Effort)

In a week or so, there is another brainstorming meeting. Steve opens up with "Well, guys, we have articulated the values and the modes of doing Advanced Technology Development. I would like us to focus today on the types of development effort that we could do – given who we are and where we want to go".

"You mean do we focus on developing some unique process technology, or instead drive development of a novel design? Or something along those lines?" asks Dr Cz.

"Well. Kind of" responds Steve. "Given that we do not have a fab or a pilot line to experiment with some new process module – what *do* we do? What degrees of freedom do we really have?"

"Hmm. Seems to me that the days of inventing new process modules are mostly behind us – even if we did have a pilot line. That was the proverbial low hanging fruit that was picked years ago. Unless we venture to the rarified heights of the real bleeding edge… Come to think of it, another thing we should peg is the time horizon of this enterprise. After all – are we doing technology development for something that could be in a product in a year or two, or in ten or twenty years. Mainstream technologies are so sophisticated and complex now that developing something disruptive and truly new – like a new transistor architecture – takes 10 years and is a huge effort" says Dr Cz, scratching his beard. "Look at things like EUV or Metal Gates or FinFETs (*TBB 1.5). The industry has been playing with those for decades. With Advanced Technology in our business, you can't just jump in and expect short term benefits."

"True dat" says Dave. "I seem to remember seeing first papers on FinFETs somewhere around the time when we were still at JKK Inc. and working on – let's see – the 130 nm node. And it does not look like it will get into production until 14 nm, at best. So… 130, 90, 65, 45, 32/28, 25/20, 14" he says counting technology generations on his fingers." That is six generations – or about 10 to 12 years. And I am sure people were working on it for a few years before publishing those initial papers".

"Excellent point" says Steve, writing 'Time Horizon' on the board, "developing some disruptive technology nowadays does take a decade or more. And it is so

R. Radojcic, *Managing More-than-Moore Integration Technology Development*, https://doi.org/10.1007/978-3-319-92701-5_6

expensive that companies tend to collaborate with the R&D Consortia like IMEC or SEMATECH (*TBB 1.6). But are there opportunities – lower hanging fruit, so to speak – to do things with a shorter fuse. Maybe in packaging or something?"

Doug raises his hand and, when everybody looks at him, says "Isn't that the bailiwick of the various suppliers in the supply chain? Foundries develop transistors, OSATS develop packages, EDA companies develop design tools, and so on. Everyone sticks to their core competence, and we are just the end user who does IC architecture and design… and maybe test and product engineering."

"Yes, good point Doug. We *are* the users and the integrators," says Steve, "but does that mean that the only advanced technology that we can do is on things that are in our own sandbox, so to speak? Are our only opportunities to differentiate our product through architecture and design? Are we doomed to use the vanilla technologies offered by the supply chain like everybody else?" and writes 'differentiation (thru design)' on the whiteboard. "Or are there things we can do outside our sandbox?"

Technical Background Box 1.5: Recent Disruptive Technologies and Incubation Period

- *Extreme UltraViolet* (EUV) is the next-generation lithography technology using an extreme ultraviolet (EUV) light source with wavelength of 13.5 nm. EUV became a leading contender to replace optical lithography in the mid-1990's (vs. X-rays, e-beam, etc.). Since then it has been consistently predicted that it will intersect volume manufacturing about 5 years from the date of each forecast, starting in ~2002. Now, high-volume use is expected by 2018–2020 for the 7 nm or 5 nm nodes. The change from the current lithography, based on 193 nm wavelength source, is very disruptive and requires not only a new light source but an entirely different optical system (reflective vs. refractive), chemistry of photoresists and developers, mask technology, inspection procedures, etc. This ongoing schedule slip in EUV deployment has been accommodated by the industry by extending the 193 nm lithography – through use of immersion litho, double and quadruple patterning, and other "tricks" – leading to an explosion of mask count (up to 100 at 10 nm node) and manufacturing cost (litho steps are repeated multiple times). Incubation period *~20+ years*.
- *Fin Field Effect Transistor* (FinFET) is a transistor where the gate is wrapped around three sides of the channel – by etching raised fins in Silicon wafers. The approach enables superior control of the electrostatic properties of the MOSFET, resulting in much better performance and leakage characteristics than those of traditional planar CMOS devices. The term FinFET was coined in 2001, and Intel first published its version – called tri-gate transistor – in 2002, but the early development effort started circa ~1998 at the latest. FinFET architecture has been adopted in the mainstream technologies at about 14 nm node – circa 2015. Incubation period *~15+ years*.

(continued)

Technical Background Box 1.5 (continued)

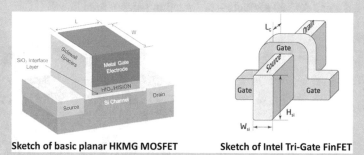

Sketch of basic planar HKMG MOSFET   **Sketch of Intel Tri-Gate FinFET**

- *High-k metal gate* (HKMG) – a planar MOSFET device using hafnium-based gate dielectric with a metal gate (typically titanium based), instead of the traditional SiO_2/SiN dielectric with polysilicon electrode. In order to enhance device performance and manage leakage currents, the industry has abandoned use of pure SiO_2 gate dielectrics circa early 1990s and has been using various types of oxynitride films since then – with progressively higher dielectric constant (k) values. Various alternative materials have been explored in mid- to late 1990s, and the initial papers about hafnium-based dielectrics appeared in the literature circa ~1996. Intel announced the deployment of hafnium-based high-k dielectrics in conjunction with a metallic gate in early 2007. IBM announced plans to transition to high-k materials, also hafnium-based at about the same time. Incubation period ~*10+ years.*

Note: Material in the gray boxes is intended for those who are interested in more semiconductor technology and/or industry background information – and may be skipped by those who are not.

"Well, yes," says Dr Cz, "there is the art and science of integration. And as the process technology matures, more and more of the value is in that integration, rather than some specialty process module. In the design space, you could for example use plain vanilla design flow offered by a selected big EDA vendor, or go and buy the best-of-breed design tools from whoever makes them, possibly build a private tool or two of your own and integrate them into a custom design flow that is optimized for your specific product. You end up with a better design flow, which ultimately allows you to design better ICs. The value *is* the integration – but the price you have to pay is to develop a methodology of your own, make an in-house point tool if needed, and write all sorts of scripts to make the tools work with each other… Presumably there is a similar opportunity in the process technology space. You obviously cannot do mix-and-match at the level of process modules – but you could mix-and-match various process flows and technology types? You could, for example, trade-off number of metal levels vs. chip size, or choose to use cheaper Silicon technologies with sophisticated Packaging technology to get some desired attribute – or something along those lines."

Technical Background Box 1.6: Some Collaborative R&D Consortia

- *Interuniversity MicroElectronics Center* (IMEC), founded in 1984, is now one of the leading organizations for R&D in nanoelectronics and digital technologies. The institute employs around 3500 researchers from more than 75 countries and has numerous facilities dedicated to research and development around the world, including 12,000 square meters of cleanroom capacity for semiconductor processing. Headquartered in Belgium (https://en.wikipedia.org/wiki/IMEC).
- *Semiconductor Manufacturing Technology* (SEMATECH), founded in 1987 and headquartered in Albany, NY (after a move from Austin, TX), was a not-for-profit consortium – funded by member dues – that performed research and development to advance chip manufacturing. SEMATECH had broad engagement with various sectors of the R&D community, including chipmakers, equipment and material suppliers, universities, research institutes, and government partners. It has been mostly wound down since 2015 (https://en.wikipedia.org/wiki/SEMATECH).
- *Laboratoire d'électronique des technologies de l'information* (LETI): established in 1967 in Grenoble, France, the institute now employs 1700 people and hosts 200+ collaborators from its research and industrial partners. It has extensive facilities for micro- and nanotechnology research, including 200 mm and 300 mm fabrication lines, 11,000m^2 of cleanroom space, and laboratories and equipment that provide first-class nanoscale characterization research capabilities (https://en.wikipedia.org/wiki/CEA-Leti).
- *Institute of Microelectronics* (IME), a research institute founded in 1991 in Singapore, focused on research in advanced engineering including integrated circuits design, Silicon process technologies, advanced packaging, and reliability testing. The institute employs ~ 350 professionals and has 14,000 square feet cleanroom, including an advanced packaging lab (https://www.a-star.edu.sg/ime).
- *Industrial Technology Research Institute* (ITRI): founded in 1973 in Taiwan and now one of the world's leading technology R&D institutions employing ~6000 people. ITRI has played a vital role in transforming Taiwan's industries from labor-intensive into innovation-driven. Over the years, ITRI has incubated over 300 innovative companies, including well-known names such as UMC and TSMC (https://en.wikipedia.org/wiki/Industrial_Technology_Research_Institute).

Note: Material in the gray boxes is intended for those who are interested in more semiconductor technology and/or industry background information – and may be skipped by those who are not.

"Maybe, something like putting several die in a single package – which may produce more bang for the buck at the system level, than pushing the SoC to the latest Silicon technology generation while leaving the support die in a separate package? Or something like that?" adds Panchali, thinking aloud.

"I like it" says Steve and writes 'technology integration' on the board.

"I would think that this is especially so the wider that we cast our net, so to speak" comments Dave. "Let me guess, an optimum implementation considering just the chip may be different than an optimum configuration considering the chip plus the package plus the PCB plus the system, and so forth. Cost of things like system test, or the cost of some other peripheral technology option in the system, may dictate a different sweet spot than what you would see if you look just at the chip cost. And finding that new sweet spot may involve a whole lot of analyses and exploration… Trial and error to get to the wholistic optimum."

"Excellent" says Steve and writes 'global vs. local optimum.'

"So, if you assume that there are, say three, sources for each technology, and you are just looking at Silicon foundries – you have a choice of three options. Simple," adds Doug.

Hmm, he spoke up without raising his hand, I am thinking. Is he getting into this brainstorming or just loosening up, and more in touch with his Hawaiian half?

"If you look at the Silicon and the Package then you have nine possible choices," Doug continues, "since you could take the Silicon from any one of the three foundries to any one of the 3 OSATS. But if you then add in all other degrees of freedom – materials, substrates, test, etc. – you end up with a huge number of options. And presumably, there is only one magic mix that is the optimum for a product. Product Delivery teams just do not have the time to go down every path and explore every combination – with their backs to the wall to meet the target tape out schedule."

"Ha! So, to draw an analogy to the Oregon Trail pioneers from the old western movies," says Dr Cz grinning, "you need a native scout to explore the landscape, before the wagon train comes through. I am thinking that leading a wagon train is sort of analogous to designing an IC – a big enterprise with lots of things going on, many moving parts, different factors to consider, etcetera. So Advanced Technology could be an equivalent to the native scout who tells the trail boss – 'no go south, big river, tall mountain, go north, flat plain, then west, good hunting'" – putting on some kind of fake native Indian accent. "So an Advanced Technology team exploring options is like a native scout exploring the landscape and finding the best path for a wagon train."

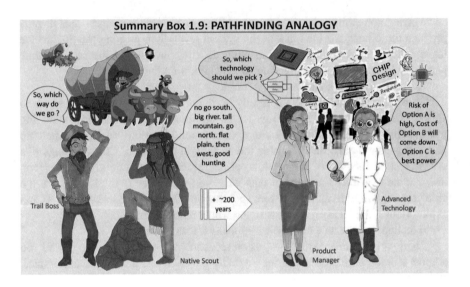

"That is cool" says Steve and writes 'PathFinding' on the board. "It is complementary to the ideas of Monitoring the Industry and Fast Failing that we talked about the last time. PathFinding would be doing the studies and the analyses – sort of trial product design – to find the best combination of technology options for a given product – but ahead of the product. Like Cz's scouts exploring the various paths, before the wagon train comes through."

"Also, if one factors in the time scale and considers what each one of those technologies may, or may not, be a few years from now – it becomes a really huge matrix of choices. How does one pick the right path – not for the current product using current technologies, but for the future product to be implemented in technologies that will be available a few years down the road?" adds Dave.

"Especially if you need to tweak the chip architecture for a given future technology mix. I mean, what if the choice of an optimum path is dependent on a specific design implementation – and vice versa – if there is an optimum design for a given selected technology path" muses Steve, haltingly, as if realizing something just then. "RoCo needs to know the possible technology options early enough to intercept its internal design effort. Having the foresight into future technology capabilities and options would be huge" he concludes. You can see that he is getting excited now – pacing the room and fidgeting with the markers.

"Furthermore," he adds, "given that RoCo is now big enough to influence the development efforts of its suppliers – rather than being just a passive user – an internal advanced engineering effort is needed to coordinate the technology development of the various partners in the supply chain… All early enough to intersect the design of our product. That would be really huge! I am sure that many of the supply chain companies would love to partner with RoCo to get its inputs on what it would like to see a generation or two down the road. This would reduce their risks – investing in technology features that are already baked into RoCo's products." And he writes 'Proactive Steering & Coordinating Partner TD' on the board.

"Now we are cooking," he adds, "time for another one of those tables that Cz likes so much" And he starts building a table like this:

ADVANCED TECHNOLOGY ~~CONSTRAINTS~~ 'KNOBS'

~~Constraint~~ Variable	~~Mode 1~~ *Go low*	*Go high*
Time Horizon	*Short – Impact next product gen* • Low risk and benefit • Low investment (not much that can be done with short fuse)	*Long – Impact n + x technology* • High risk and benefit • Hi investment (need the best in class capability)
Technology Domain	*Restricted – Internal sandbox* • Low risk and benefit • Med investment (to be best & differentiate vs competition)	*Unrestricted – Whole supply chain* • Med risk and benefit • High investment (but shared with supply chain partners)
Optimization Span	*Narrow – Focus on local optimum* • Med risk and benefit (but easy to monetize) • Lower investment	*Broad – Focus on global optimum* • High risk and benefit (hard to monetize) • Higher investment
Operating Mode	*Do everything alone* • High risk and benefit (come up with something differentiating) • Higher investment	*Partner with supply chain* • Med risk and benefit (come up with something that is shared) • Lower investment

"What do you mean with the monetizing bit in the Span row?" asks Dr Cz.

"Well," says Steve turning around and facing us, "you can do some development engineering and come up with a proprietary something that makes your product better. Presumably you can then charge more for that product, or capture a larger market share, or have a higher margin, or something, since all the value is contained within your own product. In that scenario the value is relatively easily monetized. Alternatively, you could work across the supply chain and come up with some configuration that results in some global optimum – a better system, rather than a better IC. Like, for instance, Panchali's example of putting multiple die in a single package to make a more compact, and possibly higher performance, module. The incremental value is at the system level. But the component itself may in fact be more expensive. It is not necessarily obvious that our customer will be willing – or able – to pass on that incremental system value by paying more for the component. It could be that he – or she – is not willing to pay more than the price of the sum of the unintegrated pieces. In that scenario the value is harder to monetize."

"Ah so so so. Just because something is better at the system level, it may not necessarily be easily translated to higher unit price at the component level…" says Dave. "Like for example, we could make our chips support, say, faster charging time of a phone, but at a higher component cost. It is not clear that the system vendor would be willing to pay more for a chip just to get faster charge up time" he adds, sort of thinking aloud.

"In fact" adds Dr Cz, "this bit about monetization is a good point. Since we are a Fabless entity, it is not necessarily obvious that we can monetize any learning in the process domain. Realizing process learning would require some change in the foundry's process, and they may, or may not, be willing to implement it. Or they may want to give it to all their clients – thereby erasing our opportunity to monetize the differentiation. Or they may want to charge an arm and a leg for a customized process flow, or…".

"Errr," I clear my throat, not sure whether I should comment and risk seeming stupid or add a thought and participate. "My dad – a garage manager – says that the only way he can make money off some new equipment is if it helps him do a job *faster*. He says that having some new tool does not let him raise his prices due to the competition down the road, or increase his market, since most customers like using local garages. So – the only way he gets to monetize an investment in tools is if it lets him complete a job quicker, and therefore decrease his labor cost. Wouldn't something similar apply here? In the sense that the best way of monetizing advanced technology effort may be if it reduces our cost structure – allowing for better margin. Rather than hoping to be able to raise the component price by providing increased product performance."

"Yes!" Steve exclaims "excellent point Jasmine. The link between an advanced technology investment – especially in process domain – and the ability of a Fabless entity to monetize it is not simple. If a company already dominates a given market, then increasing market share is not possible and increasing the price may not be feasible – due to competition. So, it may very well be that the only way to monetize an advanced technology investment is through increased margins. In which case advanced technology effort should focus on reducing the costs rather than improving the performance. Entirely different kettle of fish from the engineering point of view. Clearly this whole path to monetizing an advanced technology development effort needs to be thought through a lot more".

Phew… No one laughed at me. I am feeling good!

"I am thinking that the last three rows in your table all talk more or less to the same thing" adds Panchali, "you either try to do something narrow – alone or with a partner – and you end up with a local optimum that differentiates your product. And you can presumably monetize that somehow. Or you engage in a wider collaboration with multiple partners across the supply chain, and you end up with a global optimum – derived presumably through better integration. And then you may or may not necessarily be able to monetize the value?"

"True – that is the difference between the two columns – the 'go low' vs. 'go high'," responds Steve, "although there are other variants. For example, you could choose to work with the entire supply chain, but do it through a series of one-on-one agreements, always focusing on a narrow local optimum. Or you could partner with multiple entities simultaneously to find a global optimum. Or some mix in between."

He then paces a bit, staring at the board, and adds "Hmm. Come to think of it – maybe using the table format is a wrong way of presenting the degrees of freedom we are talking about. Maybe we should show it as we talk about it – a set of separate knobs which a company can dial up to its liking, as it sets up its Advanced Technology Initiatives. Something like this," and he gets up and draws a cartoon like this on the board:

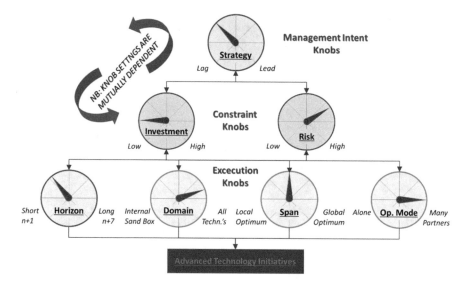

We all go quiet while he scribbles on the board and admire his sketch.

"Cool!" says Dr Cz. "I like it a lot. Not only does it look a lot cooler than a plain ol' table, but it actually does convey the message that there are a set of inter-dependent knobs which we tune to define the Technology Development Program that matches the constraints and the objectives. I like!"

We all nod.

"The different knob settings may also have different implications for the legal side of things" adds Dr Cz. "For example, if you do everything internally, then claiming the IP is relatively easy. If you partner with multiple entities, the assign-ment of the ownership of IP becomes very awkward. The lawyers would go crazy. There could be implications on ownership of our entire IP portfolio, once you factor in various derivative ideas, and so on."

"Good point about the IP" says Steve. "In fact, for some things we may want to give away the IP – even if we invent it. In some scenarios, differentiated technology may have a negative value. Being Fabless, it may be best for the company when everyone – including our competitors – uses the same process technology, even if we invented it. The reduced cost derived from such mainstreaming may be more important for the company than the differentiation obtained from having a proprie-tary process. In that kind of scenario, I guess, the company would want to give away the IP. The value of the invention would be then just a time lead – that we get to use a given technology option before our competitors do. So, sharing the IP with a supplier, or even letting the supplier own all the IP, may be the right thing for the company. It may be easier to manage the privileged access through some kind of a contractual agreement on joint development, rather than through straight IP ownership".

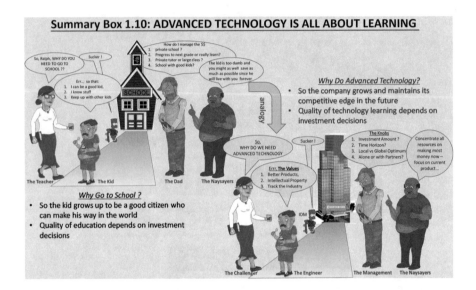

Summary Box 1.10: ADVANCED TECHNOLOGY IS ALL ABOUT LEARNING

"Bah! This is getting complicated and we are beginning to sound like lawyers. The trade-off between the business and technical value propositions... Always a dirty dark art form. Calls for Machiavellian scheming and Solomonial wisdom" adds Dr Cz. He does seem to like history, I am thinking. I mean – who talks about Machiavelli and Solomon in high tech?

Steve again takes a snapshot of the whiteboard and says "guys, this was excellent. Thank you. Many different aspects and implications in this whole business of setting up an Advanced Technology program. So, thank you. Keep thinking. In the meanwhile, I will see if I can compile some consistent message for Behrooz." And he gathers up his notebook and laptop and hurries off, while the rest of us migrate slowly back to the offices and labs.

Later that day I am mulling over the discussions and events of the day. It is after hours – 9:00 pm – and the cubicle maze is unusually quiet and almost cozy. No one is around, the offices and cubes are darkened, and there is a sense of tranquility. I stayed late to tabulate and sort through some of my lab data. And while doing this, I let my mind wander. I am realizing that there is a lot more to this advanced technology proposition than I initially thought. I started out thinking 'advanced technology = good.' Duh. Self-evident truth that pushing the technology envelope is a valuable enterprise that has to be good for a company – especially in this, so-called, 'high tech' industry. But now I realize that this was a very superficial view. No one laughed at my comment in the meeting – I am still glowing from that – but in fact, I was naive. There are so many nuances and levels to an advanced technology proposition, and the value to a company like RoCo is not necessarily obvious. It really does merit a lot of thinking through. Even grayheads like Steve and Cz are struggling with it. If RoCo does go down that path, it will be like a giant corporate experiment – which may fail. Would be great to see all that through, though.

Normally I would have enjoyed talking through some of this with Victor – he often does have an interesting way of looking at things. But he has had his head up his butt recently. No – that is not fair. He has been very excited about his new position and is busy packing and finding a new place to live in Davis and meeting all sorts of new people. I don't want to be bitchy – it is not supposed to be just about me. But, then again – recently it hasn't been about me even a tiny little bit. So, here I am, working late, rather than out on a date. Is this just a phase – or are we drifting apart? We will see. This relationship stuff is complicated. Maybe also a giant experiment – like RoCo's advanced technology? Requires investment and involves risks, and there are no guarantees. No matter how much you think and worry, ultimately you make some decisions on bases of beliefs, rather than pure cold data. I think that in the corporate world, it is called strategic intent – judging from some article I read in a magazine. And in personal relationship I guess it could be those pesky 'feelings.' It is complicated.

I think I have figured out that the first step is to know what *I* want – but, I am now realizing that, to do that, I first must know who I really am. What trade-offs am I willing to make? What am I willing to give up in order to get something else I want? Do I want a relationship with Victor enough to give up an opportunity at RoCo? Or vice versa – do I want to ride the RoCo opportunity enough to give up Victor? Am I one of those dedicated people who get their fulfillment from their profession and are willing to give up a relationship for it? Or am I a girl who values a relationship – assuming it is the right one – at the cost of a career opportunity? I am just not sure – at least not yet. Feeling a bit like a confused girl right now.

Chapter 7
The Proposals

Another few weeks go by. With Dave's help, I have managed to extract a model for the device I was testing in the lab. The 'model' is really a set of complex equations that describe the behavior of a given type of a transistor, with something like 100 different coefficients that have to be tuned so that the model predictions match the measured data. The format of the model, i.e., the basic equations without these coefficients, is standardized – fortunately. Otherwise this would have been really hard. We are using BSIM models (*TBB 1.7). Then I had to run a set of SPICE simulations – these are circuit-level computer simulations – to ensure that a selected set of the so-called Figure-of-Merit circuits behave well in simulations using the models I extracted. We found some problems – I am told that this is normal and not because my models were bad – and I had to go back and tune some of the coefficients. I must say, I enjoyed all this very much. Somehow very elegant – this describing the complex physics of a transistor that I characterized in the lab through a set of compact mathematical equations. And, I can do it! Dave likes to say that there are maybe 100 people on the planet who do this model extraction and fitting – and I have just become one of them. Yaaas!

© Springer International Publishing AG, part of Springer Nature 2019
R. Radojcic, *Managing More-than-Moore Integration Technology Development*,
https://doi.org/10.1007/978-3-319-92701-5_7

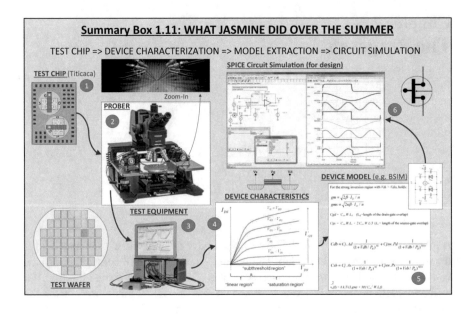

I also spent a couple of weekends with Victor. He is settled in at Davis now, and I went to see him and stayed in his new digs. It was great spending time with him and exploring the new area together. I guess it is nice up there, too. Much greener than what I am used to and quite hot in the summers. But maybe a bit too quiet for an LA girl – although San Francisco is sufficiently close to get a quick fix of that urban feeling. And of course, we talked about the future. We both have less than a year to go to graduation – hopefully – and then some hard decisions have to be made. Things cannot possibly stay the way they are now. Victor is pretty much set with his job at PetGen, so he wants me to move up with him. He said that he understood that I am reluctant to move up just for an open-ended relationship and that therefore between now and then, we should probably go through the process and meet the families, do a formal engagement, and then get married.

Just like that. He is right, of course, and I am all for rational thinking and planning – but! Really?! Come on! He suggested marriage purely as an item in a list of steps necessary to get me to move up to Davis. It *is* nice that he recognizes my reluctance to move, I guess. But, if this was a proposal, it was not what I expected. It is supposed to be a big deal – and certainly every girl is brought up to expect the angels to sing on such a magical moment. I know that he is trying to be understanding, but he did not even come out and ask me outright whether I wanted to marry him. Let alone allowing for some romance to creep into it. The question was implied, but still – a girl likes to be asked. I was… well… shook. So, I said… nothing. I don't think that Victor has ever seen me at a loss for words – so I presume that he thought I was overcome with emotions or something. I don't know. And I am actually not that sure myself what I was feeling. So, after a bit of an uncomfortable silence… we changed the subject. We went out for dinner, or something – but the elephant was in the room, so to speak. Fortunately, I was due to fly back that evening, so we never really got to talk it through. I still don't quite know what to feel or think – let alone how to respond to Victor. Besides, I got quite busy at RoCo. I have only a few weeks of my internship left, and I need to finish the models, write a QA report and file the release forms, and so on – all before the semester starts and I go back to prof Dracula's dungeons. I'll think about it later… when I have the time.

Technical Background Box 1.7: Device Simulations and Modeling

- *Simulation Program with Integrated Circuit Emphasis* (SPICE): (Simulation Program with Integrated Circuit Emphasis): a program for general-purpose electronic circuit simulation, taking a netlist that describes the circuit elements (transistors, resistors, capacitors, etc.) and their connections as an input, translating this into algebraic equations to be solved, and outputting DC voltage/current level and/or AC voltage/current vs. time. SPICE is an absolutely essential tool for transistor level design and verification and is used as a gold-brick standard in design. There are many commercial versions of SPICE (e.g., HSPICE by Synopsys, PSPICE by Cadence, etc.).

- *Device models*: SPICE simulations are fueled by models that describe the behavior of the various circuit elements (R, L, C, diode, transmission line, etc.), including compact behavioral models of transistors. The most common models are: (there are many others)

  - BSIM (Berkeley short-channel IGFET Model) for MOS devices (there are multiple versions)
  - Gummel–Poon model for Bipolar devices

  The model *format* is standardized and controlled by a Compact Model Council – an industry body currently under the Si2 umbrella (https://www. si2.org/). Full device models, including the extracted coefficients, are typically provided by the foundries – for every transistor type and every process technology.

Note: Material in the gray boxes is intended for those who are interested in more semiconductor technology and/or industry background information – and may be skipped by those who are not.

Then, an e-mail comes from Steve, calling another all-hands meeting. We all pile into the large conference room, and Steve begins with "All right guys. I have compiled all the inputs from the discussions we had over the last few weeks and now want to share what I have. Please note that this is still work-in-progress, so feel free to comment, and by all means, do bring up any new thoughts. I heard from Behrooz's admin that there will be a meeting sometime early next week, and I expect that this is where he will want to review all our proposals. Anyways, the intent of the meeting today is a sort of a dry run. Note that these are slides for The Management and are therefore suitably abstracted," he makes the air quotes with his fingers and ominously lowers his voice when mentioning 'The Management'.

He first goes through several introductory slides outlining the overall state of the industry and technology, talking about the projected slowing in the Moore's Law evolution, the rising cost of fabs, and the associated engineering efforts and pointing out that, judging from the annual reports, large semiconductor companies typically spend somewhere between 10% and 20% of their revenues on research and development. He also comments that what may be included in this R&D spending is probably different for IDM's and Fabless entities, may vary in scope, and is a variable that is likely dependent on the specific accounting approaches. He closes that portion of the presentation emphasizing that he is proposing a much more modest approach for RoCo, at least to begin with.

He then puts up a slide that says:

Proposed Advanced Technology Initiative: the Values

- **Track Technology Development**
  - o   3 to 5 year 'Headlights' (n+2 node and beyond)
  - o   No Surprises - from Potential Disruptive Changes
- **Optimization of Our Architecture & Design**
  - o   Timely Feed-Forward to Product Teams
  - o   Differentiated Infrastructure (e.g. Co-Design)
- **Generate Technology Integration IP**
  - o   Defensive and/or Offensive Opportunities
  - o   Chip-Package-PCB Technology Integration

RoCo Inc

The points he makes on this, and the backup slides, are similar to the discussions we had during the brainstorming sessions. He emphasizes each point by verbally elaborating it and adding illustrative examples of past disruptive surprises, past opportunities for differentiating the design infrastructure, and past examples of chip-package-board integration opportunities.

Interesting… If I was making the slides, I would have written down everything – ending up with a lot of text and subtext and notes and things. His slides are so simple and clean but convey all the information – especially with his talking through the examples. I guess that is 'management-speak'. I will try to do something like this at the U from now on. I am also enjoying the fact that I have participated in a process that is now bubbling up to the company CEO. Cool!

Steve then goes on to propose approaches for an advanced technology effort at RoCo, again using a main slide with several backups and again supporting each point verbally with illustrative examples. He sounds so convincing and competent. I am impressed. The key slide was:

Proposed Advanced Technology Initiative: the Approach

- **Engage with Consortia**
  - Participate in Collaborative Pre-Competitive TD
  - Explore and Encourage the Best Fit for RoCo
- **Engage with Academia**
  - Contribute to R&D of the Leading Profs
  - Insights into R&D + Access to new Talent
- **Partner with the Supply Chain**
  - Steer their Technology Development (e.g. JDP)
  - Focus on Integration and Design

RoCo Inc

This is something that we have not addressed directly in our brainstorming sessions, but given all the conversations, and his presentation, I follow the intent. Basically, he is proposing that we engage with the consortia and academia as a way of tracking and influencing the long-term technology trends – in a more direct way than by just participating in the various industry conferences and forums. This would give us the 'headlights' he spoke about earlier, as well as getting a seat at the table, so to speak, that defines future technologies. And he is proposing that we engage in collaborative engineering projects with select partners in the supply chain in order to coordinate and drive their short-term technology development efforts to suit RoCo's needs.

It is interesting to see it all condensed onto a single slide – that even I can get. Ha!

Then Steve moves on to suggest mechanisms for managing and controlling the proposed advanced technology efforts, using:

> **Proposed Advanced Technology Initiative: the Management**
> - **Set up a Strategic Steering Committee**
>   - ○ Senior Management Guidance
>   - ○ Link to Strategic and/or Market Roadmaps
> - **Define and Manage a New Project 'Funnel'**
>   - ○ Propose and Review Potential TD Projects
>   - ○ Implement 'Feed-It-or-Kill-It' gates
> - **Define Technology Release Level System**
>   - ○ Track Technology Maturity (R&D to Production)
>   - ○ Phased Transfer to Product Delivery Teams
>
> *RoCo Inc*

This is again something we have not talked about, but Steve goes on to explain that the company needs some mechanism to steer any advanced technology effort, because it is new for us, and the management needs the visibility to control the costs and the investments. He emphasizes the need to connect market strategy with technology development to ensure its convergence with other strategic intents. He also elaborates that RoCo engineering culture is very much focused on product development and that an advanced engineering activity that is not tightly anchored to a specific product would wither and die – unless the management maintains the focus and drives the strategic vision. With the second bullet, he makes the point that Advanced Technology Initiative should be an ongoing effort and that, rather than trying to identify specific projects to invest in right now, we should define a mechanism for reviewing and dispositioning such project proposals on ongoing bases. Verbally he proposes potential candidates – such as design for manufacturability, or chip-package co-design – but mostly for illustration purposes. Interesting that the examples he selected are the ones within our 'sandbox', as we talked about. And with the third bullet, he makes the point that as we look further into the future – as we extend the 'technology runway', to use his phrase – we need to define a set of milestone criteria by which we move a given technology through various stages of development in a structured and disciplined way, from concept through demonstration and then all the way to production. He also talks about the need to involve product teams at various points in the evolution – not too early, since they are usually very busy but not so late that they cannot impact the development effort.

Very clever, I thought. He is focusing on setting up the mechanisms necessary for managing the proposed Advanced Technology effort, rather than on specific projects. It is not like anyone would disagree with the proposed mechanisms – allowing for a clean consensus. Whereas different people may have different favorite projects and different perceptions of opportunities – leading to discussion and dissention. Clever. Also, he keeps coming back to the point about the need to stay connected with the product teams – I guess to make an advanced technology activity palatable to a product-centric culture. A lot for me to learn here.

And then Steve addresses some of the key issues and concerns, with this slide:

Proposed Advanced Technology Initiative: the Challenges

- **Quantifying the ROI**
  - Long gap Between 'Investment" and 'Returns'
  - Many hands touching it
- **Monetizing the Value**
  - Value = Learning & Fast Failing ($$ later…)
  - Local vs. Global Optimum
- **Transferring to Product Teams**
  - Need a mechanism for graceful hand-off
  - Harmonizing with Multi-Sourcing

RoCo Inc

He elaborated on the challenges of monetizing the value and also emphasized the issues – new to me – about the difficulty of integrating some new and potentially differentiated technology with our standard sourcing practices. I think that this refers to the discussions about the advantages and disadvantages to a Fabless entity of having a differentiated process within a distributed supply chain.

And then Steve closes with a few slides that address the projected budget requirements. He proposes a multiphased program and focuses on the first exploratory phase. He proposes that the initial budget for Phase 1 Advanced Technology Initiative include a staff of up to 5 people for a few months, along with an incremental travel and expense budget. Basically, he is requesting management authorization and staffing just for Phase 1, to conduct a more in-depth assessment of the practicality of an Advanced Technology Initiative by a Fabless company like RoCo – by reaching out and talking with the various potential partners and suppliers. And he closes with a proposal to come back in a quarter, or so, with a revised overall proposal, as well as a set of specific projects that has been vetted by a steering committee, and a better budget estimate, based on more concrete information.

"And that is it" says Steve, turning off the projector. "What do you guys think?"

Everybody is digesting it, so most of us look at each other and do not say much.

"Well, I think it is excellent" says Dr Cz, "you outlined an approach – basically focusing on partnering with the existing supply chain – and defined the potential values and risks. You are asking for a small investment of time and treasure for Phase 1, in order to get to be more specific for Phase 2 and on. That way you are not asking the management to commit to a complete program – something that is not fully defined yet – a proverbial 'cat-in-the-bag' so to speak. You are deferring the definition of specific engineering projects, but at the same time, you are asking for their philosophical blessing. Very elegant. Get them pregnant with the concept of Advanced Technology Initiative now, and tap them for larger budgets later. Asking for much more at this stage would be scary for them, and trying to move much faster than what you are asking for is probably unrealistic."

The rest of us all nod in general agreement. I don't know about everybody else, but personally – I am stoked! It was very exciting for me to see all this – up close and personal – from concept to management proposal. Rad!

"Well, thank you. And thank you all for helping me put it together. Wish me luck next week. I will let you all know how it goes" says Steve, packs up his staff and walks out.

Chapter 8
The Good-Bye

Today is my last day as an intern at RoCo.

And yowza! What an experience! I had a great time, learned a lot, met a bunch of nice people – and got paid for it! I am now familiar with device lab setups and know how to extract a model – something that Dave calls a baseline 'fulcrum point' for everything in IC design. And I witnessed the start of an Advanced Technology experiment at RoCo. Steve and Dr Cz have been on the road quite a lot recently – presumably meeting with people and gauging the support from potential partners, as per Steve's slides. I guess that this must mean that Steve got the green light from the management – at least for Phase 1. I think they said that last week they were in Belgium to meet with IMEC. And they mentioned stopping in New York and talking with SEMATECH. And they returned from Taiwan and Korea a couple of weeks ago. I do hope that the whole thing does get off the ground by the time I am done with grad work and ready for a full-time job. Would that not be just perfect for me! Well, a girl can hope.

I have been so busy here, trying to wrap everything up, that I have not had much of a chance to ponder the Victor thing. Funny, that. Normally I would think that a relationship with Victor, and maybe even a marriage, is something that is far more important than a device model at RoCo. Ergo, a rational person, that I claim to be, should be thinking about that. And if I was more like mom meant me to be – girlie and romantic and all that – I should be swooning and dreaming about the wedding. But instead I am working 12 hours a day, 7 days a week – not so much because I have to, but more because I want to. I wonder if that is a telling sign? Seems like I am more excited about device modeling and RoCo Advanced Technology than I am about living with Victor. Or is it that I am just reluctant – maybe scared – to embark on a big change like moving and marrying? And am therefore just practicing the art of procrastination? At least for now. I mean, I do intend to marry and have a family – someday. Not doing that is just not an option for me – having babies is something that I have always taken for granted. Maybe just not yet? Or is it that I am not feeling 'it' with Victor? I do love being with him, but it is not supposed to be like this. Somehow – maybe the way I was brought up – I expected that a prince charming

© Springer International Publishing AG, part of Springer Nature 2019
R. Radojcic, *Managing More-than-Moore Integration Technology Development*,
https://doi.org/10.1007/978-3-319-92701-5_8

would come along and sweep me off my feet with angels singing, birds chirping, and rays of sunshine peeking through the clouds. And I would be transported to the seventh level of fulfillment – without feeling the need to rationalize anything. The romantic crap that they feed you – Disney and fairy tales and moms. A girl can be happy only once she has got her man. Really!? But then again – I am not buying the line about not needing a man. Ultimately, a husband, a family, babies – it is what it is all about for me... Maybe just not yet? Or maybe just not Victor?

"Ready?" says Doug, standing in the 'door' of my cubicle. The team is taking me out for an extended lunch at Duffy's – a good-bye treat is apparently a tradition. They tell me that once, long long ago, they did not take an intern out for a good-bye treat and the poor guy never graduated and turned into a pile of dust – or something equally dreadful. They say that they do not wish to have me on their conscience and that therefore they must take me out. Some of them – Cz, Dave, Doug... – have been working together a long time and have evolved a history and traditions. Nice.

"Oooh. I did not realize it's that time already. Yes yes, I am coming," I say turning off my computer. "I got to stop in the bathroom – so why don't I just meet you guys there," I say.

"OK," says Doug. "See you there. Don't be long or you will miss the first round."

I need to change my shoes, double check my make up, and make sure I am presentable. Today – on the account of the special occasion – I decided to forgo my usual jeans and tennies for something more formal: slacks, a blouse, jacket, and heels. I decided that wearing a dress would be going too far for these geeks. I give myself a critical once-over in the mirror, straighten my jacket, and walk over to Duffy's. Nice day outside. Birds are chirping, and San Diego sun is shining.

"And there she is," yells Dr Cz, when I enter Duffy's and walk over to their table.

"Who is this pretty young lady? I am not sure I recognize her," asks Steve, as Doug pours me a beer.

I sit down and see that they already have a pitcher of beer and a plate of the usual finger foods. Steve, Cz, Doug, Dave, Panchali... everyone is here. Even Walter – our lab tech – is here. Walter Chu normally hides in the lab and does not like to socialize. Apparently, he wanted to be an artist, but his parents – good tiger-mom types from mainland China – made him go into engineering. So, he has a full EE degree, but according to Doug, he finds the non-exempt 9:00 to 5:00 lab tech job attractive because it allows him the time and energy to pursue his art passion.

"So, now that you know us a bit better, are you ready to come to work for RoCo," asks Steve. "I mean after you graduate, of course."

"Well, now that you know me a bit better, are you ready to offer me a job," I shoot back. "Assuming that I do graduate."

"No worries about that," pipes in Dr Cz. "It is unanimous – you are good. And I spoke with prof Dracula, and you could be done in a couple of semesters."

"Yeah, Jasmine must be good. I understand that Doug allows her in the lab unattended," comments Dave. "This is a rare honor. Doug normally guards his lab like a rabid dog, and no one – not even god – is allowed in without his or Walter's supervision."

"Well…," says Doug. "You can never be too careful. But Jasmine – she has the right level of respect for the lab aumakua. Very disciplined. She says a prayer before getting on a prober, respectfully thanks the instruments when she is done, and makes sure that she never turns her back to the lab altar. To Jasmine," he responds joking and raises his glass.

"Oh, so that mumbling she does while she is running SPICE is a recitation of her mantras? Well it seems to work – her model fit is unusually robust for something produced by a newbie," teases Dave. "To Jasmine."

"And I must say, I love the way she applies the wisdom of her dad's garage to our industry. Excellent insight, there," adds Steve. "To Jasmine."

And they all raise their glasses. I like it – but am embarrassed too. "Errr. Thanks guys. I had the best teachers," I respond – and hope not to be blushing too much.

"See," says Dr Cz. "We all want you back. Don't even bother to look at the offers from other companies. They are all bad, anyways. I am told that TekMOS guys eat baby-whales for breakfast," he jokes.

"Especially if things go as we expect" adds Steve "this Advanced Technology Initiative is looking quite promising. If all goes to plan, we will have a job not just for you, but for all your fellow grads. And more. And it will be really fun work. We will have our fingers in many pies, so to speak. Excellent learning opportunities for us all."

"So, your meetings with Behrooz went well?" I ask.

"Well, that went as well as can be expected," he responds. "But the thing that I am really stoked about is the reaction we are getting from the people in the industry, in academia… everyone we talked with. Seems like participation in cross-industry activities by Fabless companies is something that has been kicked about at various high-level industry forums. SIA, GSA, and the like (*TBB 1.8). Apparently, a number of big wigs have talked about a need for a more active role by the Fabless sector – especially in advanced technology and integration activities. We seem to have lucked out in terms of timing... This is a hot topic, and supposedly people from outside the company have been whispering in Behrooz's ear about it. RoCo could be a leader in this space – a fabless entity participating in shaping the technology and the industry roadmap. It is all very exciting and looking very promising."

"What about the foundries and the OSATs? Are they eager to have our hands in their pants," asks Dave. "Normally they are quite protective of their turf."

"Actually, they all are very interested to have us be much more involved than we are now. Not just to define the technology requirements, but also to participate in the actual technology development. They want an understanding end-user in the cockpit with them, so to speak – participating in the various tradeoff decisions that inevitably come up. Some of them even mentioned hands-on collaboration in the design and characterization of the technology test chips, and the like," responds Steve. "That would be a really intimate collaboration in Technology Development."

"And it is not that they want us just for the extra engineering resources that we would bring to the table," adds Dr Cz. "They are all only too aware of the challenges with new technologies and the slowing of the Moore's law cadence, and are looking for pro-active ways to ensure that their process technologies are easily integrated with design, packaging and other elements of the product sourcing eco system. So, they are looking for partnerships with an integrator like us. They want us for our pretty pointed heads – not just our hands."

Technical Background Box 1.8: Major Semiconductor Industry Bodies and Forums

- *Semiconductor Industry Association* (SIA), founded in 1977, is a trade association and lobbying group that represents, and is the voice of, the US semiconductor industry. One of the main achievements of SIA was the creation of the first National Technology Roadmap for Semiconductors, in the early 1990s – which then evolved into ITRS (https://www.semiconductors.org).
- *Global Semiconductor Alliance* (GSA), founded in 1994 as a Fabless Semiconductor Association (FSA), is an organization that hosts multiple councils and forums intended to promote and facilitate the workings of the distributed supply chain for the semiconductor industry (https://www.gsa-global.org).
- *Institute of Electrical and Electronics Engineers* (IEEE) is a professional association formed in 1963, whose objectives are the educational and technical advancement of electrical and electronic engineering, telecommunications, computer engineering, and allied disciplines. It hosts many conferences, forums, publications, and standards bodies (https://www.ieee.org).
- *Silicon Integration Initiative* (*Si2*), founded in 1988, is a nonprofit consortium of semiconductor, systems, EDA, and manufacturing companies focused on improving the way integrated circuits are designed and manufactured, with primary emphases on standards for design and design-process interface (https://www.si2.org).
- *Joint Electron Device Engineering Council* (JEDEC), founded in 1958, is an independent semiconductor engineering trade organization and standardization body, with activities focused on standardization of memory interface formats, part numbers, electrostatic discharge (ESD), reliability and lead-free manufacturing (https://www.jedec.org).

Note: Material in the gray boxes is intended for those who are interested in more semiconductor technology and/or industry background information – and may be skipped by those who are not.

"And the Consortia and Academia?" asks Doug. "like IMEC or the UC profs."

"Well, they do want to find ways to tap the money from the Fabless sector. Traditionally their support came from the IDM's, but with the Fabless sector becoming such a large portion of the semiconductor industry they are struggling to find ways to access this pool of funding. So, they are very eager to have us join and are all trying to figure out ways to make their learning more attractive and accessible to Fabless entities," responds Steve.

"Cool," I say. "Congratulations. And thank you so much for allowing me to participate. It has been an eye-opening experience for me. Thanks for involving a novice like me."

"We need the fresh blood," says Dr Cz in a funny accent – trying to sound like Count Dracula from Sesame Street, I think. He speaks with a bit of a funny accent anyways and should give up trying to act out others – he is so bad at it. "So, getting you all primed and ready to deliver is in our interest. It wasn't just because of your pretty face, you know," he continues.

"Well, we do need to bring up our quota of pretty faces around here, what with the likes of yours Cz. Or yours, Dave," retorts Steve.

"Hey, hey," pipes in Panchali. "What about me. I am not pretty enough for you? What are you suggesting about me and my looks? You saying that I am ugly and old, or what?" she demands in a mock aggressive tone.

"Ok Steve. Need some help in pulling that foot out of your mouth," teases Dr Cz.

And so it goes. A very fine, fun afternoon. I was touched. I felt like I was a part of the team. Not sure that I have earned it – yet – but it felt good. We stayed, joking and talking, munching and sipping, until it was time for my exit interview with Human Resources. Then, I had to say good-bye to them all and leave. I managed to not cry – but it was close.

"Keep in touch, Jasmine," yells Dr Cz, as I was walking out "and remember – we are expecting you back."

Chapter 9
a Technologist: Cz's Contemplations (Circa Year 0)

I am a lucky man…. I stumbled into this career of mine – and wow – what a ride it has been! I studied Electronics at the university, all those years ago, because my dad always said that it was good to have a 'craft'. And I did not think that modern history or political sciences – the things that I really liked – was what he had in mind. Not good 'bread', to borrow another one of his phrases. So, almost randomly – I picked Electronics and Solid-State Physics. I did like Physical Chemistry in high school, and altogether math and sciences came easily to me. But, truth be said, at the time, I was far more interested in girls, hanging with friends, and other such profound things, rather than anything to do with sciences. I was not like my bro – who was born to be an engineer – tinkering with model boats ever since he was a kid. But I lucked out. I picked Electronics – and it seems to have suited me. Quite abstract and clean – no greasy dirty machines, no noisy construction sites, no bricks and concrete. The downside of a profession like Semiconductor Technologist is that it is hard to explain to the regular people – it is too abstract and does not involve things that are easily seen or touched. In fact, I have long ago given up trying. Even my kids are not sure what I do and choose to believe that I work for Mafia instead. However, ever since I finished my PhD – which, again, truth be said, I did mostly to avoid army conscription and to prolong my student days – I never had any troubles finding work. In fact, I was very fortunate and always had jobs that were custom-made for me, as opposed to having to fit in some predefined role. I have worked in a fab, in a lab, and in front and back offices. I have worked in a large vertical company, in a medium-sized EDA company, in a small startup, in a fabless company, and as a one-man consultant, servicing clients ranging from startups to financial institutions. And I had positions as a hands-on engineer, as a people manager, a project manager, a team manager, a customer manager, and a business manager. Truth be said, this breath of experiences was not a result of careful career planning on my part – more a case of stumbling into this or that. And the work has been an adventure. Throughout

© Springer International Publishing AG, part of Springer Nature 2019
R. Radojcic, *Managing More-than-Moore Integration Technology Development*,
https://doi.org/10.1007/978-3-319-92701-5_9

my career I relished going to work, day in and day out. There was always something interesting and exciting going on there. I do feel for the people locked into jobs that they hate. Must be awful. For me, participating in the semiconductor industry, and contributing to the development of the Si Technology – arguably the greatest exponent of human intellect – has truly been a privilege. My contribution is infinitesimally small – minute – but the overall enterprise was huge. So, I was like a tiny ant who helped build a great ant hill. And I even got paid to do it – fed my family, put my kids through college, and, barring any unforeseen disasters, will likely enjoy a relatively comfortable retirement. A lucky man, I am, indeed.

And – another way in which I lucked out – my whole career has been with a single technology type: Silicon technology. I sometimes wonder what it was like for the previous generation of engineers – the guys who were brought up on vacuum tube technology and had to learn solid-state Silicon technology in mid-career. It must have been hard to absorb a technological change that was radical enough to obsolete much of their knowledge and experience. Back to school in their 40s or 50s. And it will be hard for Jasmine and the next generation of engineers – because Silicon technology, as we know it, is coming to an end. I don't know exactly when, or if it will be replaced by carbon nanotubes, or graphene, or quantum dots, or god knows what else. But I do know that the single paradigm that has spanned my entire career is coming to an end. That paradigm – typically described by Moore's law on the process end and Von Neumann machine on the architecture end – is running out of gas.

Silicon technology, and the arc that it followed for the last half century, has been truly awesome – resulting in the ubiquitous electronics and so many seemingly fantastical appliances, from calculators and computers to music players, phones, Internet, and … everything else. It seems that it is now impossible to settle even the simplest of transactions – like figuring out the change for a $5 bill used to pay for cup of coffee, without reliance on electronics. And it is totally ludicrous to suggest that once upon a time a man was shot to the moon – and brought safely back – mostly using slide rule technology. Unthinkable!

Altogether, the changes in our lives enabled by the evolution of Silicon technology have been amazing – even to someone like me. For example, I distinctly remember thinking that the marketing buzz that came up when the mobile phones first appeared – some 10 to 15 years ago – was a bunch of hooey. Something like 'call person to person – anytime – anywhere on the planet'. Seemed silly at the time – most people on the planet did not have access to a phone, and even for those who did, what with time zone differences and managing the access to land lines, you had to schedule calls ahead of the time and pay dearly for the privilege. But it has happened: person to person, anytime and anywhere on the planet, is here and now. And then there is that stereotypical sci-fi vision of an omnipotent oracle that holds all of humanity's knowledge – like in that scene in *The Time Machine* where the protagonist finds some central repository of knowledge that describes what happened to mankind over the eons. That seemed to belong in sci-fi until quite recently. But it too has happened. In fact – the reality is even better than the sci-fi version – we have access to pretty much all of humanity's knowledge in our pockets. And so on. Seeming miracles abound. All made possible by Si technology.

And that technology, in itself, also seems miraculous. I am an insider and work on the technology on daily bases – but am still amazed and impressed with some of the statistics. Billions of transistors per chip and each and every one of them working correctly to a spec controlled to +/– few mV's … Feature sizes of a few nanometers – with manufacturing process control specifications that are managed to +/– an atom or two... Material purities of +/– 1 part per billion – where only one out of billion atoms is allowed to be of a wrong kind, or in a wrong place... And all that in routine mass manufacturing, producing acres and acres of working Silicon chips every month. Wow!

But making it all happen involves 'tricks' that feel like violations of the laws of physics. For example, defining features on a chip that are only a tenth of the wavelength of light that is used to print them, or storing only a handful of electrons in a potential well to make a memory bit, do feel like a violation of the basic laws of physics. In fact, the number of technology solutions that only a few years ago I said would be impossible, and that are now in mass manufacturing, is downright embarrassing. Goes to show what I know. The industry has demonstrated an impressive track record of creating miraculous solutions for various seemingly insurmountable technical challenges. Like pulling rabbits from a magic hat. As long as there was money to be made. And the money is still there, of course. There are no signs of saturation in demand, or of mankind wanting fewer electronic gizmos.

But that trajectory of the semiconductor technology – and everything that it has enabled – is coming to an end. The industry will have to absorb some kind of a paradigm shift in order to continue feeding the demand for more integrated, smaller, and cheaper electronic devices.

Si technology has progressed and grown in complexity to the point where it now 'feels' like a Rube Goldberg type of a system – a very complex, inter-dependent set of mechanisms to do something that used to be simple.

Simple Way to Take Your Own Picture by Rube Goldberg

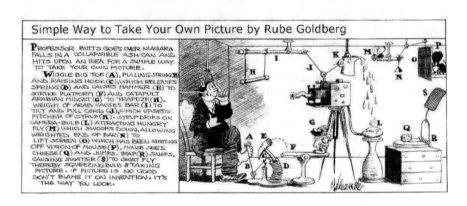

Automatic Garage Door Opener by Rube Goldberg

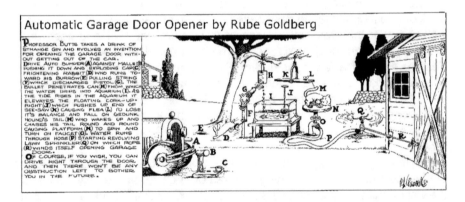

Silicon technology differs from the Rubes in the sense that all the complexity is actually required and is vitally necessary, and the end result is anything but simple. But modern Silicon technology – like the Rubes – does consist of layers upon layers of technical solutions, each invented to solve some constraint, stacked on top of each other, and ending up with an extremely complex and fragile system. A house of cards. Pushing it further feels like it is going past the limits of reasonableness – getting too complicated. Like we are pushing it beyond its natural and elegant limits – as if forcing round pegs in square holes. The technology is amazing but pushing it further along the same path seems a bit crazy.

Consider, for example, the lithography process – the printing steps used to transfer a design pattern onto Silicon wafers. This is a key process module that dictates the minimum feature sizes – and hence the overall density of transistors on a chip – and that dominates the overall cost – since it is repeated multiple times on every wafer. In order to realize features much smaller than the wavelength of light used in the optical equipment currently available, the design has to be doctored. A whole bunch of constructs must be added to the masks, in order to compensate for the light interference patterns projected on the wafer and to print the desired shapes. The pattern on the masks no longer resembles the original design, or the resulting pattern on the wafers, but is instead generated through very complex mathematical transformations to 'trick' the laws of optics. Crazy!

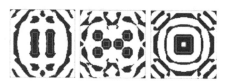

Patterns on the Masks -
shown in red - required to
realize the desired polygons -
shown in yellow

And when that becomes insufficient – sometime beyond the current 28 nm node – the design will have to be broken down and printed on wafers in multiple passes, each one printing alternate features, resulting in very complex constraints on the shapes and sizes that can be designed. All this because the wavelength of light used in the litho equipment is 193 nm – and the features that need to be printed are 22 nm or smaller. A tenth of the wavelength of light! Crazy!

And the grand solution for these extreme Resolution Enhancement Techniques (RET), and double, or eventually quadruple exposure, is the so-called extreme ultraviolet (EUV) litho, with 13.5 nm wavelength light. Use of light with shorter wavelength seems obvious, but the way that this light is generated is using machines so large that the factory has to be built around them. And the way that they work is that a droplet of tin is dripped into a vacuum chamber, and zapped with a laser, so that it explodes and radiates this specific frequency of light – which is then collected and reflected off the masks onto wafers. And to be production worthy, this must be repeated precisely and exactly a few bazillion times for every die on every wafer, and on 1000's of wafers per month. Crazy!

Consider the thickness of the films used in construction of the chips. Even the small applied voltages used in modern ICs produce electric fields that are extreme enough to trigger the bizarre quantum effects like electrons tunneling through barriers – not only as a parasitic effect but an intended mechanism used in operation of some devices. Only a few years ago, these phenomena were the exclusive realm of thought experiments – like the famous Schrodinger cat. And now it is in production. Crazy! The electric fields are so extreme that the films are at a verge of being literally torn apart, requiring thickness control that is measured and managed to atomic dimensions. And the virtually perfect materials used are only good enough to postpone this inevitable self-destruction to just beyond the specified device life – maybe a few years nowadays. Crazy!

Consider the MOS transistor itself – the heart of the Si technology. Its performance – basically the time that it takes to switch from a '1-state' to a '0-state' – is ultimately limited by the time it takes a few electrons to go from one end of the device to the other end. For years the magic of Moore's law was enabled by making the transistors smaller – thereby reducing the distance that the electrons need to travel. That was simple. But when this became too hard, and various laws of physics got in the way, the electrons were made to go faster. Crazy! And the way that they are made to go faster is to replace just the right number of Si atoms, in just the right places, by just the right atoms that are a bit larger than the Silicon atom, so that the crystal lattice is stretched in the local area by just the right amount so that the electrons can slip between the atoms that much easier. Crazy! And this is done to billions of transistors on every chip, on thousands of chips on every wafer, and 1000's of wafers every month. Crazy!

Consider the other end of the technology spectrum – the design. Operating frequencies of multiple GHz, resulting in transistors switching from 1's to 0's or vice versa, in periods of less than a nanosecond, are a norm that is routinely achieved in PCs, phones, and many other electronic gizmos. A nanosecond is short enough that the distance traveled by light in that time – at that ultimate 'absolute' of Einstein's relativity physics, the speed of light – is comparable to the dimensions of the systems: a few cm's. So, a neighboring chip on a PCB that is a few millimeters away is considered to be a long way away – and for many applications, it is too far away. Crazy! And in order to cope with this, the architecture of the system has evolved to multiple layers of memory banks, placed closer and closer to the actual processor. And the content of these memories has to be managed by shuffling data – controlled by ever more complex software – to ensure that the exact data required by the processor is pre-loaded into these memory banks at just the right time. A few nanoseconds late is too late. Crazy!

And with transistors switching 1's and 0's a few billion times per second, the power density consumed in regions of a chip is comparable to that of the surface of the sun. It is a small amount of total power – but with features being as small as they are, the power *density* is huge. So that the challenge of sucking away the heat, so that the chip does not melt itself, has become an absolute critical limit of the technology. Heat sinks, heat pipes, water cooling, and all sorts of plumbing are essential for high tech. Crazy!

And with billions of transistors on a chip, all doing their thing, the total power consumed by the high-end chips – however they are cooled – is getting into the region where the power can be reasonably expressed in units of horsepowers. Crazy!

And so on. Examples of technical solutions that seem to have gone beyond the bounds of common sense are numerous and mind-boggling.

So, the current implementation of a leading-edge Si technology does involve layers upon layers of ever more complex solutions and tricks and techniques – all necessary to address the various underlying challenges. And the technology *is* pushing the limits of laws of physics – with features approaching atomic sizes and heat fluxes approaching the un-coolable. Silicon technology has lost its intrinsic elegance – and pushing it much further is getting progressively more expensive and more far-fetched. So, I do believe that the arc that has been followed by the industry throughout my career is coming to an end. This arc – usually referred to as 'Moore's Law', but in fact involving so much more than just cramming more transistors on a chip at a reasonable cost – cannot be pushed much further.

I am an optimist and have no doubt that the industry will come up with solutions that will allow it to continue putting ever more information at our fingertips, in ever smaller, more connected, and more ubiquitous devices. But it will just have to find technologies alternative to semiconducting Silicon – sooner or later. Thus, there is a paradigm shift coming our way, somewhere around the corner. I don't know when, or what – but I do believe that it is coming.

I don't believe that this paradigm shift will happen overnight – or any time soon – since right now there are no credible alternatives to the trusty Silicon technology. But I do believe that it will happen during Jasmine's career. She is a smart kid and quite level-headed – so I am sure that she will cope – but somewhere in her career she will have to learn an entirely new technology that is completely and totally different than the Silicon technology. It will probably be a gradual shift, where Silicon technology is extended to meet the mainstream requirements, and this new technology – whatever it is – is applied only to some niche application. And, over time, the niche will grow and grow – until ultimately Silicon technology either becomes as obsolete as vacuum tubes are now or is relegated to applications that suit it more naturally. No more pushing Si technology to ever crazier extremes.

Of course, between now and then, our Silicon technology, and the associated ecosystem, will continue to evolve and morph – but in ways that deviate from the past trajectory. It is not any more just a matter of scaling feature sizes and rail voltages and cramming more transistors per chip. We will have to solve the on-Silicon challenges like the litho printability issues, of course. But we will also have to invent new materials and processes; we will have to address the multi-physics challenges like thermal or mechanical stress management; we will have to deal with integration at the system level and develop some type of co-design methodologies; we will need to develop memories that are based on something other than charge storage; we will need to invent architectures that rely on more parallelism rather than more speed; and so on. Especially the new integration methodologies. With Moore's law paradigm drawing to a close, the so-called More-than-Moore paradigm (*TBB 1.9) will evolve and be the way to getting more functionality in a system, rather than just more transistors on a die. New integration technologies will be the name of the game for a while… Lots of new things that will have to be developed and deployed, even before the big paradigm shift comes along. And most of them will be off the traditional technology arc.

Technical Background Box 1.9: More-Moore and More-Than-Moore

It is obvious that Moore's law cannot be extended forever just through dimensional scaling – sooner or later the dimensions necessarily reach intrinsic atomic limits which cannot be pushed further. Consequently, starting at about 90 nm node in mid-2000s, the industry had to abandon the traditional 'happy scaling,' based on just the direct reduction in dimensions and voltages, according to the original precepts of Moore's law and its corollary, as expressed through 'Dennard' scaling (dictating that applied voltages must be reduced commensurately with reducing the dimensions in order to manage power and electric fields – see https://en.wikipedia.org/wiki/Dennard_scaling). At that time two paths were identified for continuation of the price-performance improvements associated with Moore's law (Moore's law is basically an economic law dictating that the cost, and hence price, per chip would remain about constant when doubling the transistor count while halving the area of the features). These two paths were named, somewhat tongue in cheek, as 'More-Moore' and 'More-than-Moore' paths, as was illustrated in the 2005 issue of the ITRS:

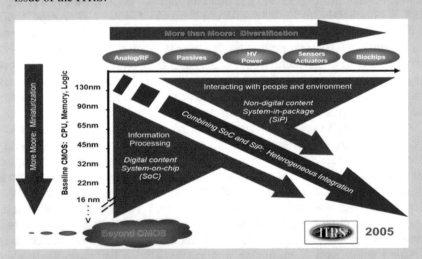

- *More-Moore* (MM): This path focused on continued increase of the *integration at the chip level*, by leveraging mechanisms alternative to simple dimensional scaling to achieve the desired cost-power-performance

(continued)

Technical Background Box 1.9 (continued)

benefits of each successive Si technology generation. New phenomena, such as strain engineering, have to be evoked to enhance device performance without necessarily reducing their dimensions. New materials, such as high-k dielectrics and metal gates have to be deployed to manage the transistor leakage mechanisms associated with the extremely low dimensions. New materials and processes, such as low-k dielectrics and air gaps in the metal stack, need to be developed to reduce the capacitive coupling between the wires, to manage the interconnect performance. New process technologies, such as immersion lithography, double/quadruple patterning, and eventually EUV, must be developed to realize manufacturability of smaller features. New device architecture, such as FinFETs, has to be embraced to manage the power-performance trade-offs. And so on – enabling the continuation of cost-power-performance benefits of successive Si technology generations. Note that following the 28 nm node, the rate of improvement realized is lower than was traditionally possible – but it is still there. However, this trajectory is reaching its limits, as exemplified by the end of ITRS – the last report was issued in 2015, and even the ITRS 2.0 effort has been wound down in 2016.

- *More-than-Moore* (MtM): This path focused on continued increase of the *integration at the system level,* by leveraging alternative technologies to achieve the desired cost, form factor, functionality, and power-performance of end appliances. The intent of this approach was to leverage more than just the digital CMOS technologies, such as the rf/analog, power management, sensors, MEMS, and other novel devices, and to combine those using new integration technologies, to produce results similar as those delivered in the past through Moore's law scaling. That is, the objective is to cram more functionality at the system level, rather than just cramming more transistors per chip, through use of advanced packaging and system integration technologies. This path leverages new 'System-in-Package' (SiP) integration technologies – as opposed to the 'System-on-Chip' (SoC) technologies – such as technologies for stacking die on top of each other (3D integration), placing multiple die in a single package (2.5D integration), etc. More-than-Moore technology path is not fully and precisely defined as yet, but this term applies generically to the class of technologies that are focused on package and system level integration (see, e.g., 'More-than-Moore 2.5D and 3D SiP Integration' by Riko Radojcic, 2017, ISBN 978–3–319-52,547-1).

Note: Material in the gray boxes is intended for those who are interested in more semiconductor technology and/or industry background information – and may be skipped by those who are not.

To draw an analogy, to me it feels like for the last 30 to 40 years the industry has been driving at full speed down a predefined autobahn. But now it is dark, and we are off that autobahn. So, we better put on the high beams, and look carefully ahead, and hope that we can spot the curve in the road. Especially since we know that there is a curve somewhere ahead. Somewhere… Requires different driving skills.

So, I believe that we will need to hone our lateral thinking skills and learn to reach out across the traditional interdisciplinary barriers and borrow knowledge from new arenas in order to come up with new types of solutions. That is, I don't think that the future is just a matter of digging deeper in existing trenches – more a matter of reaching across the trenches.

Given this perspective, I got to believe that a company like RoCo – or for that matter, any company that plans to be around for a few more years and has the resources beyond the hand-to-mouth operations – must dedicate some effort to look ahead for the technology changes that are coming. A portion of the engineering pool must be let out of the trenches, released from the pressures of the next tape out, and allowed to look further out. It is not clear how tightly coupled to product realities these extra resources need to be. Too loose a connection – and there is a risk of creating a research ivory tower – and arguably there is enough of those. Too tight a connection – and there is a risk of engineering myopia and lack of innovation. Figuring out the right horizon, and managing engineers without developing a product, is a challenge – corresponding management skills need to be developed and honed. In the short term – this or next year – the bean counters must be prepared for these resources not producing anything tangible; they will be an additional cost item with no impact on the corporate top line. But in the long term, they may help steer a company along a tricky path and will help it to avoid disastrous crashes, or obsolescence. Not doing that would be like sticking our heads in the sand, so to speak. After all, if not a company like RoCo, then who? And if not now, then when?

Good Way to Run a Company   **Not Good Way to Run a Company**

I know that I have often been accused of wearing rose-colored glasses. I know. In fact, I choose to do so. So, I *am* very excited about the Advanced Technology Initiative and the development program that Steve is pushing. It is definitely the right thing to do – for the industry, for the profession, for the company, and for us – and especially for young engineers like Jasmine. It will help develop exactly the right kind of skills and thinking necessary to deal with the future. And it may produce a technology solution which will carry us for the next one or two steps. I feel like we have a moral responsibility to drive this type of advanced development and the thinking process that goes with it...

Part II
I, Engineer (and Running Integration Tech Development)

This section, set a year later, finds Jasmine – now graduated with a PhD degree – joining RoCo semiconductor company as a full-time engineer. She joins the same group where she interned and finds the team embarking on an aggressive path to define and deploy an Advanced Technology Integration program. The team struggles with defining the mechanisms for managing the program while maintaining the momentum and the management support within an organization that is focused on shorter-term challenges. Jasmine discovers that the hard part – having decided what she wants – is making the trade-offs among the various competing priorities and desires.

Chapter 10
I'm Back!

Today marks the end of my first week as a full-time engineer at RoCo Inc.....
Tadaaaa!

It has been almost exactly a year since I left. And now, a year and a PhD degree later, I am back. And what a year it has been. The defining event was, of course, completing my PhD and officially becoming a *Dr.* Jasmine Lopez! Yaaas! I did it! But, now that it is behind me – it was a bit of an anticlimax. You do all this work, thinking, anticipating, dreading, planning … and focus all your energy on this one thing – in my case it was completing the thesis – and when it actually happens, it is a bit of a letdown. It may be just the case of what Prof Dracula calls 'post PhD blues'. Apparently, it is quite normal to feel a bit depressed after graduation – something to do with the absence of the adrenaline that you got used to pumping in order to get there. Sort of like new moms and postpartum blues.

I must say that he – Prof Dracula – has been fantastic. The man is fire. He pushes his grad students hard, gets them to do all this extra work, never seems to be quite satisfied, appears to be ratcheting up the requirements on daily bases, and all the other things that earned him his moniker – but when it comes to publishing the work, pulling together the thesis, and getting it reviewed and approved, he has been super. Thinking back, I now understand why he did the things he did, and I am glad that he has done them. Thanks to him, completing the thesis and graduating were almost easy. It was more or less just a matter of stapling together all the various project reports and papers that he made me do over the last few years. I now love the guy. I hope that I have managed to express my appreciation adequately. I thanked him of course – many times and profusely – but still… At the graduation, all I could do is just hug him and try not to cry. The best adviser – ever.

After that final review of my thesis and sitting through the oral cross-examination, the graduation itself was … well … just a ceremony. But, what a ceremony. Hundreds of people packed in the auditorium to witness the almost rote pomp and circumstance, made formal with all the faculty displaying their ceremonial robes, friends and family in suits and ties, and of course, all the grads in the rented robes and funny hats. Most of it was listening to the droning voices reading names, waiting

© Springer International Publishing AG, part of Springer Nature 2019
R. Radojcic, *Managing More-than-Moore Integration Technology Development*,
https://doi.org/10.1007/978-3-319-92701-5_10

for your turn – and then – it's over in a second. The voice calls 'Jasmine Lopez'…
walk on stage… shake hands… smile for a photo… and it's done! Still – it was
special. And afterwards, watching all the grads and families milling around was
somehow touching. Take those rich Asian brats out of their shiny Beamers, and they
are pretty much the same as the nerdy Anglos out of their beat-up K-cars – proudly
posing for pics and trying not to be embarrassed by the families. And yet, somehow,
they – I should say 'we' – swallow our 'I-am-too-cool-for-this' attitude and treasure
the moment, maybe as much as the families do.

I went to the ceremony mostly for my parents' sake. Probably much like 95% of
the kids… errr... grads there. The least I could do for mom and dad after all the support they have given me – both emotional and financial. And, I must admit it was
great to see them there.

Mom was… well a few inches taller. She has always been a black sheep of the
family, so to speak, and having her snooty Iranian cousins watch her daughter getting a PhD – in engineering, to boot – was somehow a vindication for her. She has
a funny family history full of love-hate relationships. Her mom, my grandma, was a
second wife of this rich dude in Shiraz – as old as the kids of his first wife and all
living together. This, supposedly, was not unusual in Iran – especially in those days.
The first wife, and her children, apparently never liked her much – but, when my
grandma died while still quite young, it fell on her half-siblings to bring up my
mom. Mom was never close to them, not just in age but emotionally too, and seems
to have related to them more as distant aunts and uncles than as brothers and sisters.
It seems that the family business was going well, and she was sent abroad for her
education – partially to get her out of the family's hair, supposedly. And when the
revolution happened – she just stayed. She hooked up with dad and dropped out of
college and went to an art commune instead. Dropping out of school was bad, but
marrying a non-Iranian, and a Catholic, was the final straw, and the Iranian family
cut her off. So, she never went back – although she has stayed in touch with some
of them. I actually know little about that side of the family – other than the stories
that she has told me. Some of the children of her half-siblings have subsequently
moved to California and have reached out to us – but we somehow never got close
with them. Pretty much just getting together for occasional Persian New Year and
other formal events. However, traces of the Iranian upbringing keep wafting back
into mom's life – so she gets uptight about the appearances, and what the rest of the
family may think, and the right ways to behave in front of them, and so on. So, with
her Iranian hat on, so to speak, she thinks that my PhD is some serious status – and
so she invited the Iranian cousins to my graduation – mostly to show off. I think that
she sees it as a kind of justification of the choices she has made. Whatever. So, on
top of being happy for me, and proud for herself, she was also pleased to… well…
thumb her nose at her Iranian family. Or that is how I read it.

And dad – well he just could not stop grinning – in spite of being stuffed in a suit.
I know he hates that. He is a second-generation Latino and grew up used to California
casual. So, he is normally much more at home in his garage overalls or in his dopey
'lucky-Hawaiian shirts' that he sports when watching 'futbol' with the neighbors,
than in a suit. He apparently tried to 'join the man' as he puts it and wore shirts and

ties when he went to college. But he dropped out after 3 years of Mech Eng. He concluded that his real love was cars and that he did not need a degree to do what he liked – customized rides and restored primo vintage cars. I think that it was also that he met mom – they fell crazy in love – and dropped out together. And it was the 1970s. They must have been high. And since then he never tires of gloating about how rarely he has to wear a suit – used to be just for marriages, baptisms, and funerals – but it seems that graduations also qualify. He was much more himself – out of the suit and tie and a beer in hand – at the graduation party that he organized for after the ceremony. It was virtually a block party – with all of the family, my proverbial 'cien-Lopez-primos,' neighbors, and friends – the party spilled out of our house into the cul-de-sac that we live on. It was great. Lit. Lots of food – and beer – and talking, laughing, gossiping, and loud 1970s rock. Interesting seeing all the neighborhood friends – the kids I grew up with – somehow looking at me a bit differently. I am not sure if I am imagining it, but behind their jokes and calling me 'maestra' or 'profesora,' it felt like they were somehow more distant – more offstandish. Maybe not.

Even Dario – my bro – flew in to be there. He is with the Air Force and seems to have arranged a 4-day leave just to make it to the party. Air Force really suits him – in spite of being gone most of the time. He was always into airplanes – even when we were kids. Currently he is stationed in Germany, but before he did a tour in Guam and Korea, and he did the academy in Colorado. Although we are not very tight – it has been so since we were kids, since he was outdoorish and I was bookish – but his coming all that way just for my graduation was special. It was a big surprise – he did not tell anyone that he is coming – and both mom and I screamed and cried when he showed up. Dad just grinned even more.

I guess getting a PhD *is* a big deal – not just for me, but for the whole fam. After all they were much more than just cheerleaders – they invested time and money in me and my degree and are pleased and proud that I made it. Mom and dad dropped out of school, Dario did the academy, so going all the way to a 'Dr' is a first. I suppose electronics suits me – a nice combination of my bookish tendencies and all that bonding time I did with dad in his garage. It *is* engineering – but not the greasy, noisy, bulky kind that dad likes. Although some of his love of cars seems to have rubbed off on me. I cannot help but notice the cars that people drive and always ponder whether it suits them or not. Like other girls notice dogs and their owners – I notice cars.

One person who was not there was Victor. We – in all honesty I should say 'I' – decided to end it. It wasn't just that he moved up to Davis and that I did not want to move out of SoCal or that I was feeling like he took me for granted. It was more a matter that I came to a conclusion that our chemistry wasn't right. The strongest bonds that we had were cerebral. I mean I did like being with him, we did have a great time together, and we did make each other feel good – but the best part with him was the discussions and debates that we would get into. It was intellectual. Other parts – the sex, the togetherness, the sharing, and so on – were good, but it was the cerebral part that was great. After his so-called proposal, I got to thinking. I look at mom and dad – quite an improbable combination of an uptight Iranian artist and a laid-back Latino car geek. WTF! How does that make sense? My mom tried to bring us up with her Iranian ethic – another reflection of her conflicted background – but she for sure did not practice it in her life. At least not when she was young. For all I know, they might have hooked up to begin with due to drugs, sex, and rock-n-roll, but they have stayed together all these years, not just because of me and Dario, or the absence of the divorce option in both of their cultures, but because of that mysterious 'chemistry' that they seem to have. I believe that they are actually very happy together – improbable as it may be. I can see it in their eyes – even when they bicker and things. I want some of that, too. My relationship with Victor was more consistent with my version of mom's Iranian rules – it made sense to me because we were intellectually matched. But it did not have that passion that seems to have driven mom and dad. I want more of what they have – more of the 'did' part in mom's 'do-as-I-say-not-as-I-did' dictum. So, Victor and I talked about it and I decided that we should go our separate ways. He tried to talk me out of it, but it felt to me – maybe unsurprisingly, given my new perceptions – like he was appealing to my head and not my heart. And it was the heart that needed satisfying. It was sad in a way – but the right thing for us. We are supposedly still friends – but somehow, it is not right. Maybe the strain will go away with time, but right now, we don't seem to be able to recapture even that intellectual bond that we had. So, he did stuff with his family, and I had my great graduation party.

That party was a really nice milestone … A kind of a punctuation mark – a period at the end of that chapter of my life, so to speak. I am starting the next chapter with a totally clean slate. No school, no Victor – or a boyfriend of any kind – no strings. I am ready to get into whatever is next. So, when Dr Cz called to congratulate me on graduation and to offer me a job at RoCo, I took it! And here I am. I took a couple of months off and did pretty much nothing – hanging on the beach with friends from school, reading trashy novels, occasionally helping dad in his garage or mom in her gallery – so nothing, really. So, now I am here and truly ready.

I now report directly to Dr Cz – so I guess I was promoted. And I got an office, rather than a cubicle, with an actual door – but no window. I share it with Elvis. It is true – my office mate's name is Elvis. Apparently in China and Taiwan, they get to pick their English names in school – the teacher writes a bunch of names on the board and kids pick out the ones that they like, or that are not taken yet. So my office mate ended up being an Elvis. He is Taiwanese, did his PhD in Vanderbilt, and is now a new engineer working in Product & Test Engineering group. He started a couple of months before me. I must say I get a kick out of his mixture of accented ESL and various southern terms that he picked up in Tennessee. He is a nice guy – quiet and blissfully absent from our office – either working on the test floor or hanging out with other Chinese engineers. So, I think we will get along quite nicely.

I have of course caught up with the old squad, too. We have stayed in e-mail touch since I left, but it was nice being here and seeing them all in person again. Doing coffee with Panchali, tech-talking with Doug and Dave, admiring Walter's latest art… Dave had his Vette repainted – classic red and white. Nice! He and his car – one excellent match.

I also had a long meeting with Dr Cz – for him to bring me up to speed and to define my projects. After the casual chit-chat about the graduation, the university, the prof, and what have you, he puts up his feet on the desk – still displaying his same old worn-out boat shoes – and says:

"So, I don't need to explain to you what the group is all about and what we do – that is one of the great advantage of re-hiring an intern. Especially a good one like you," he says, smiling. "But there have been some changes, of course. Since you left, we have sharpened the proposal for Advanced Technology Initiative, and The Management has blessed the basic program. Steve has defined and structured an Advanced Technology Steering Committee, comprising of fairly senior management from several organizations, and has used this as a platform to obtain concrete funding for several specific projects. Forward-looking projects that are not tied to product delivery – therefore true, honest-to-goodness advanced technology projects …"

"Oh, that is great! So, you have gotten the approval for the Advanced Tech program that we talked about when I was an intern. Nice. Congrats. So am I to be a part of that team?" I ask.

"Well, yes and no. There is no specific 'Advanced Technology team' now," he says making the air quotes. "We have an approval for a program that is, in principle, sized to be about 20% of the overall Infrastructure Organization, plus some additional expense and travel budget."

"Infrastructure Organization? Huh?" I ask.

"That is the Technology and IP organization, as opposed to the direct Product Delivery groups that work on designing or managing the products. Steve's whole team, plus the Foundry & OSAT interface, and the library groups are all a part of it," he says. "The approved concept is to dedicate about a day-a-week of that group of people for advanced technology projects. That is a day-a-week of the engineering headcount responsible for bringing up new process technologies and for developing the basic IP. Right now, it works out to an equivalent of about 20-ish people. As opposed to a fifth of the overall RoCo engineering headcount, which would run into three digits."

"Excellent. So, can I be one of those 20," I ask, not really following – or caring – right now about the rest.

"Well, it is not managed that way," he responds. "The program is managed through assignments rather than job functions. So rather than going off and hiring a bunch of new people, the idea is to build slack in the existing organizations and borrow some time from the current headcount to drive the approved Advanced Technology Initiatives. That way, we can tap into a broad skill mix pool that we already have on board, rather than trying to squeeze all the right skills within the budgeted headcount." He pauses a bit, recrosses his legs, and continues, "So, most of us spend some of our time on Advanced Tech projects, and some on the product related projects. Some of us spend more, some spend less – depending on the momentary need of the approved Advanced Technology Initiatives, and the slack in the product related projects … By the way, we have decided to call this whole enterprise 'Advanced Technology Initiative' or ATI," he clarifies.

"ATI? … You know ATI is All Terrain Innovation – a company that makes parts for ATV's and other such things. So, are we making those extra grip soft dune-buggy tires?" I say, joking. I guess, you can take a girl out of a garage, but you cannot take the garage out of this girl.

"Anyway, how does it really work? Who gets to work on what and how is it decided" I ask, getting back to being serious. "I need someone – you, I guess - to tell me what to work on," I add.

"Understood and understandable. Don't worry – I will keep you busy and off the streets," he says grinning. "You will have more work than you can handle. The way it is managed is through the time card system. Basically, we estimate the effort required to drive a given ATI project and Steve gets the overall budget for that project approved by the Advanced Technology Steering Committee. A project lead is defined, and he – or she – goes off and recruits the engineers from existing teams to work on that project. Basically, an implementation of the classic matrix-organization structure. This involves a definition of the required engineering skills and the associated effort, finding the people with those skills, and negotiating the required time with his or her manager. Then a project team is formed, with participation that flexes with demand and availability. Some of the people on the team may be asked to spend a lot of time on that project this week, but not so much the next week, and so on... Overall, the sum-total of time spent on that ATI project should fall within the approved budget."

"Ah, I see! That way you access a whole spectrum of engineering skills… A design guy, a SPICE dude, a Test guy, and what have you, for a part of their time. I get it," I exclaim. It is beginning to make sense to me. I may be slow, but... "I suppose that the engineers like it too? They get to work on advanced things that are good for their careers – at least some of the time - rather than just working on the one thing that is their core job."

"Yup," he says. "That is the theory. So, your initial assignment will be to help design the next technology test chip. Some of the structures that you work on are required for the next product, and you will charge that time to the appropriate product charge numbers. And some of the structures will be for the ATI purposes, and you charge that time to ATI project numbers. There is a project lead for the test chip who manages the exact content, schedule and day to day activities. In the case of Poopó – no joke, that is the code name for this test chip - the lead is Brian," he concludes.

And he then goes on to explain the intent of the approved ATI projects (*TBB 2.1) and the type of test structures that need to be included on the Poopó for these projects.

Technical Background Box 2.1: Example Advanced Technology Initiative Projects

- *Design for manufacturability* (DfM) – a collective name for a set of multi-disciplinary engineering practices intended to make IC chips more readily compatible with a specific manufacturing process, thereby resulting in higher yield and faster yield ramps. DfM, typically in form of good design practices, design guidelines, and tribal knowledge, was always practiced in semiconductor technology. However, starting with the 90 nm node, or thereabouts, new systematic design-process interactions became a significant yield loss mechanism, and more formal and explicit methods had to be developed. New DfM solutions were required to address, for example, the printability issues (to cope with the sub-wavelength lithography), planarity issues (to address pattern-dependent CMP and etch-loading effects), or critical area issues (to address layout dependence of yield). This class of DfM solutions is typically implemented through a three-pronged approach: (1) The manufacturing process is characterized – typically using test chips – and specific attributes are described through a set of appropriate DfM models. This is the realm of the foundry. (2) These DfM models are then imbedded in suitable design and/or verification CAD tools. This is the realm of the EDA companies. (3) The DfM tools are then used by the designers to massage the design – typically at the polygon level – to ensure the compatibility of all layout constructs with the manufacturing process. This is the realm of the design teams and the Fabless companies. Since the failure mechanisms were due to new phenomena, and since multifaceted solutions were required, this type of DfM practice was a new specialty that most companies included under the technology development umbrella.
- *Design for variability* (DfV)– variability in the manufacturing processes (e.g., deposition temperatures, process times, power settings, etc.) and the consequent variability in feature characteristics (e.g., film thicknesses, line widths, etc.) are intrinsic in the nature of the semiconductor technologies and have always been there. Traditionally, this variability was addressed through the use of 'corner models', where parameters are set to correspond to the tails of the parametric distribution (typically $+/- 3\sigma$), such that most of the population (~99.7%) fits within the selected corner values. However, starting around the mid-2000s, many new process-design interactions resulted in increased relative variability in device characteristics (producing larger spread between the corner models), while the reduced operating voltages led to increased sensitivity to this variability (requiring smaller spread between the corner models). This then precipitated an issue with the traditional corner-based approach. Both, the number of corners and the spread between corners increased, resulting in inefficient IC design. Consequently, new practices for addressing variability – collectively

(continued)

Technical Background Box 2.1 (continued)

referred to as statistical design or design for variability – had to be evolved. These practices require new methods for characterizing the statistical process and device characteristics (by the foundry), new statistical models and tools that describe the variability (by the EDA companies), and new design methodologies to better target the IC design (by the designers). Development of these new methodologies, tools, and models also fell under the advanced technology development umbrella in most companies.

- *Design for thermal* (DfTh) – continued increase in the integration levels results in increasing power even in CMOS-based ICs, leading to increased heat dissipation and a rise in average chip temperature. Managing the average steady-state chip temperature is not a new challenge, and high-performance systems have relied on water or air cooling for decades. However, with the end of "happy scaling" period of Moore's law (where voltages were more or less scaled in proportion with the dimensions in order to maintain reasonably constant electric fields – as dictated by Dennard's law (https://en.wikipedia.org/wiki/Dennard_scaling)), the power density increased for each technology generation. This leads to increased local heat dissipation and increases the temperature of the local hot spots. Managing the temperature of local hot spots on an IC has become an issue starting around the 2010s, especially for mobile devices that do not have the luxury of active cooling systems, such as fans. Consequently, a set of new methodologies – especially new to the semiconductor world – had to be developed. This required new methods for characterizing and modeling the thermal behavior of ICs, new thermal simulation practices, and new design techniques – collectively referred to as design-for-thermal practices. All aspects of design impact the thermal performance of an IC – from chip architecture to floor planning and physical design, to package design, and even the PCB and system design. Hence design for thermal is truly a multidisciplinary activity, which also tends to be addressed as a part of technology development activities.

Note: Material in the gray boxes is intended for those who are interested in more semiconductor technology and/or industry background information – and may be skipped by those who are not.

"So, you have two roles," I end up concluding. "You are a project lead for some ATI projects and you happen to be my line manager, too. Right?" I ask.

"Yes, yes. We all wear all sorts of hats. I am a group manager, and a project manager for some of the ATI initiatives, and may be a hands-on contributor to other projects, and so forth. You are a part of my line org and will also wear several hats. Some of the things I ask you to work on will be driven by product needs and current technologies. And some of the projects that I ask you to work on will be ATI projects looking ahead to the future technologies and are independent of any specific products."

"Confusing," I say. "But I think I get it. Either way, I work on what you - with your line manager hat - tell me to. And for some projects my day-to-day project manager may be someone else – like Brian. And for some projects you may be both my line manager and my project manager. Confusing, but I think I get it."

"Yes," he says. "Let's not worry too much about it right now. It is complicated, but we can make it work -mostly because the overall Infrastructure Organization is reasonably small, and there are a several of us who know most of the people, and their background skill mix, etcetera. So, we know who to go to and tap for a given task. Let's work with it and see how it goes. Either way – there will be a lot more meetings for you to attend to clarify each and every one of the projects that you are assigned to. Project meetings, group meetings, team meetings and so on... Don't worry – you will soon get on all sorts of e-mail lists and you too will be like the rest of us – spending your days running from meeting to meeting trying to stay on top of all this, and then doing your real work at night. And speaking of meetings – I got to run." And he gets up, collects his laptop, and disappears.

So that is how it started. I am all excited and woke and very much looking forward to it all.

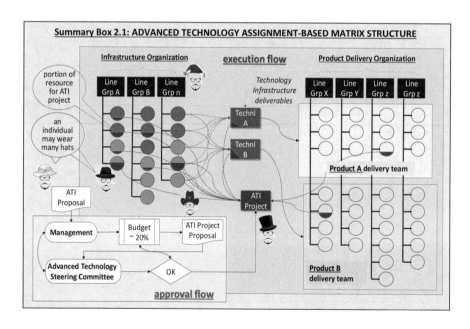

And today the team is taking me out for a 'welcome back' treat at Duffy's. It seems that there is another tradition that Dr Cz takes out his new hires when they come on board but that they get to take out the whole team if they leave. Dr Cz claims that this discourages engineers from leaving and that this is the reason for the stability of his group. He has apparently practiced it for decades. Really?! Like I am going to base my career choices on a $100 dinner treat that I may have to pay for. But I like it that they have 'traditions'.

And tomorrow – in fact this whole weekend – I am apartment hunting. Real adulting! I am temporarily rooming with friends from the U and am really looking forward to having my own place. My own place! Me … alone … by myself … my own bathroom... I can walk around naked without worrying about flatmates or their friends or whoever. My own place! Just one of the ways how I plan to spend some of that big money that RoCo is paying me now. Not to mention that suit I saw at Theory – a girl needs her clothes to look professional. And a safe parking spot for my new ride – a 1965 Mustang Convertible that dad restored for me. A graduation gift he said. Rad! Life is good, and this girl is hot!

Chapter 11
Managing Learning (in a Matrix Organization)

We have a design review of Poopó test chip today. It's a pretty big deal – a formal review of the test chip that has to be 'passed' in order to get the approval, budget, and headcount, to proceed to the next phase. There are altogether three of those: one, called 'preliminary review', to approve the overall content of the test chip before anything is designed; one, 'tape-out review', to approve the final design and layout before the chip is taped out to the foundry; and one, 'final review', to review all the deliverables, including the documentation, and often some of the initial test results. The review today is 'tape-out design review' – and is a culmination of many months of my, and several other engineers', work. This is a half-day meeting where engineers present their structures, and everybody gets to pick at them. The management – Dr Cz and Steve, his VP along with a few other engineering VPs – and all of the design engineers who worked on Poopó, and some of their managers, are here. Looks like 30-odd people.

This is the first design review for me, so I am a bit nervous. I worried about screwing up my presentation – but Brian has had several dry run meetings over the last couple of weeks, and he tells me that I will be fine. Panchali – back to being my protector mother hen – warned me that the VPs normally mess around on their laptops and tune in only occasionally and that if they are not paying attention, it is not because there is something wrong with my presentation. Then this morning I fussed about the right thing to wear: my usual jeans and top or something more formal, like slacks, jacket, and blouse? I ended up opting for my usual 'uniform' – since I decided I don't want to stick out – and it is unlikely that my fellow geek engineers own anything other than what they always wear. And here we are. The best part, of course, is that the meeting is scheduled to run through lunch – so RoCo pays for the pizza! No beer, though. Boo.

"All right, let's begin" says Brian, standing up and showing his first slide:

© Springer International Publishing AG, part of Springer Nature 2019

R. Radojcic, *Managing More-than-Moore Integration Technology Development*,

https://doi.org/10.1007/978-3-319-92701-5_11

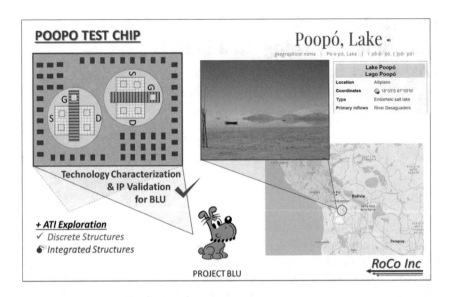

"First of all, the name of the chip is pronounced 'popo' with a long 'o.' It is named after a lake in Bolivia and not anything else that may come to mind," he says with a smirk. Everyone chuckles.

Brian is a middle-aged guy, with a freckled boyish face and a tinge of an Irish accent. Middle aged but rides this hot fire-engine-red Ducati crotch rocket to work. Midlife crises? I wonder? But he is not the type – he is very level-headed, direct, and down-to-earth. Realistic and somehow all there. I have enjoyed working with him. Seems to have a calming effect on a roomful of agitated engineers, and his body language always seems to communicate a sense of confidence in a project.

He continues "Poopó is the primary test chip for technology characterization and IP validation for Product Blu. As always, there is good news and bad news. The good news is that all IP blocks and class A structures – those are the technology deliverables for Product Blu – are on schedule and included on Poopó. At this point, there is low risk with any of these," he pauses a bit, and I am wondering who comes up with all these code names and the naming conventions for various chips. 'Blu' is a code name for the next product chip – apparently also a name of a dog in some Brazilian comic strip – adhering to the system of naming product chips after cartoon dogs. Test chips seem to be named after lakes. Was there a meeting, with all the high-powered VPs, that decided this?

Then he concludes "the bad news is that only about half of the class B structures – those required by the ATI initiatives – are going to make it on this chip." He pauses, waiting for a reaction.

"And why is that?" asks Steve, as on cue.

"Well. Two reasons," responds Brian. "Firstly, because some of the class B ATI structures are just not done. Discrete transistor arrays and other non-integrated structures are complete, but the integrated circuit structures are not. And secondly, I do not believe that we can possibly catch up. This chip is a part of a shuttle and therefore has to tape out in exactly 2 weeks – there is absolutely no slack there. So, we do not have an option to slip the schedule."

'Shuttle' is, I have been told, a program that foundries run, where they aggregate a number of test chips from multiple users onto a single mask set and process them in a single wafer run. The analogy is that it is like getting a seat on a shuttle flight. In fact, the foundry manager of the program is referred to as the 'shuttle captain.' Funny.

"Why?" asks Steve again. "Is it that we do not have sufficient resources? I thought that we had the staffing for this project – including the planned ATI structures? Can we assign incremental design and layout resources and pull the schedule in? What can be done to recover?"

Layout is the term used to describe the actual drawing of the shapes that are eventually implemented in the masks. This sounds easy – after all, all you have to do is operate a CAD tool to draw a bunch of polygons. But in fact, it is deceivingly complicated and requires specialized techs – who are apparently often in short supply. Like drafts people in architects' offices, I presume.

"Well, it is not a matter of design or layout resources. The issue is that these structures, by definition, are not normal or within the standard design rules approved by the foundry. So, we need to define a few new mask layers, change a number of design rules, upgrade a few models, build a custom PDK, design a few specialized cells, and revise the entire Design Enablement (*TBB 2.2). Without these upgrades, we cannot do the design and layout of the requested ATI circuits. The bottleneck is in the PDK group," responds Brian, patiently.

"OK," says Steve "so, can we get the resources from the PDK team?"

"Errr, I will let Fred speak to this, but basically, the PDK resources that were staffed to Poopó were pulled off, and now it is too late to catch up. Fred…?"

Fred – the manager of the PDK group – looks up from his laptop and says, a bit belligerently, "Look. We have received a series of changes to the technology spec from the foundry, and we had to update the PDK for Product Blu. I have only so much capacity, and when I have to choose between supporting an ATI initiative or a product, the product wins… every time."

Technical Background Box 2.2: Process Design Kit (PDK) and Design Enablement

PDK is a set of files that define all the models and rules that describe a specific manufacturing process – to be used by the IC design tools. PDK is essential and a must-have item for physical design and verification of EDA tools – without which IC design is just not possible. PDK is typically created and qualified by each foundry for use in a selected reference design flow but may be enhanced by a design house for use with specialty tools and/or designs. Typically, a PDK contains:

- *Design Rules* – a document that defines the layers and the associated layout conventions used for a given manufacturing process, describes the

(continued)

Technical Background Box 2.2 (continued)

basic layout constraints (e.g., allowed line widths and spaces, etc.) and includes DfM rules and guidelines for optimizing the layout

- *Design Rule Checker* (DRC) – a file used in physical verification tools to ensure that a layout complies with the design rules
- *Layout Parameter Extraction* (LPE) – a file used in physical extraction tools to convert a layout into an equivalent electrical circuit and to assign the resistance-capacitance values to given constructs, which are then used to estimate the circuit timing characteristics
- *Device Models* – a file that contains the coefficients for standard device models used for SPICE circuit simulation, including the nominal and corner models, and voltage and temperature dependencies
- *Layout vs. Schematic* (LVS) – a file used in design verification tools to ensure that the electrical connections contained in the layout are consistent with the design intent

Design Entanglement includes the PDK plus additional elements required for design, including any special EDA tools or scripts, a qualified design flow, and technology-specific items such as

- *Standard Cell Library* – a collection of basic designs that perform various fundamental logic functions used as building blocks in digital IC design. The library typically includes multiple instantiations of the same logic functionality that enable different performance vs. size trade-offs, allowing for optimization of IC price-power-performance characteristics. Standard cells are described by a variety of separate files that represent the name and symbol, layout, footprint, connectivity, and timing characteristics of each cell.
- *IP Blocks* – including memory generators and/or memory instances (typically using the bit-cell provided by the foundry) and other frequently used standard higher-level blocks.
- *PCells* – a collection of files that parametrize various layout constructs typically used in the design of custom or analog circuits, allowing for faster and better layout and design of analog blocks

Note: Material in the gray boxes is intended for those who are interested in more semiconductor technology and/or industry background information – and may be skipped by those who are not.

"But," responds Steve, "Poopó is our one chance for technology learning this year. There are no other test chips scheduled anytime soon, so if we miss this tape out we delay an opportunity to learn by many months. We were all in the Behrooz meeting that blessed the Advanced Technology Initiative, and I thought that we were all nodding up and down and agreed to support it. So…?"

"Hey, get off my back. The foundry sent us an unscheduled update, and there was nothing that I could do. Product must win," responds Fred, rather emphatically.

"Ok. I understand. In a head-to-head conflict, the product must win. But could we not have done something proactively to cope? Brian, how come you did not raise the flag earlier and alert us all in time? We just will not be able to do Advanced Technology Initiatives if we run projects in this reactive way. You are telling us now that we have no way of recovering. If we were having this meeting a few weeks ago, maybe we could have? Maybe we could have trained up someone else from another team to do the necessary PDK updates for ATI," says Steve, trying to be calm but turning a bit red in face. He is clearly getting worked up.

"No," Fred jumps in. "PDK is complicated and has to be done by one of my guys. Every time we tried to cut corners in the past, we regretted it and ended up having to redesign product chips. Millions of dollars and months in time to market. PDK engineering is not something that can be picked up overnight. If I am to release a PDK rev, it must be done by my guys," exclaims Fred, tapping on the table with his hand to emphasize the point and raising his voice a bit.

"But…," Steve responds,

"Ok Ok. Let's take this offline and have a separate meeting," cuts in Alex, their boss – a VP. "Brian, take the action to get Fred and Steve together. Please include me too. Back to Poopó, please."

The rest of us breathe a sigh of relief. The atmosphere in the room was getting tense and somewhat uncomfortable, so most of us suddenly seemed to have discovered something very interesting on our respective laptops…

So, Brian then goes on to present a series of slides that summarize the chip content and schedule and then calls on each of the designers to describe their set of structures. By the time it is my turn, the meeting is running late, and the presentations are proceeding quickly and with no comments from anyone. So, I present my structures, go through their intent and status, and… sit down. Phew. It's over and no one gave me a hard time.

Then the pizza is delivered, and whatever attention was left in the room just deflates – like a balloon. So, Brian suggests that we wind down the meeting and agree to release Poopó to tape out with whatever ATI structures make it in time. Everyone nods, eying the pizzas. He then goes over the action items – one of which was to call a separate meeting with Fred and Steve – and closes the meeting.

We all get a couple of slices of pizza, and most people wander out of the conference room. I sit down next to Dr Cz and ask him "Why is Steve picking on Brian? That was pretty harsh."

"Steve wasn't picking on Brian," says Dr Cz, seeming surprised. "He was just making a point about ATI support. Brian knows this. He was just a messenger"

"Still… Brian tried very hard to keep us all in the loop and to keep Poopó on schedule. I know that he has been camped out in front of Fred's office every morning and evening for weeks now, trying to get us the updated PDK," I say, feeling like things were not fair.

"Yes, I know. The issue here is not communications or Brian failing to manage the project. Don't you think, Steve?" he says and waves Steve over to come and join us. "Jasmine is feeling like you were too hard on Brian."

"Don't worry Jasmine, it's all in a day's work. But I will talk to him to make sure that he is OK," says Steve sitting across from us. "The issue is the organizational structure – not Brian," he continues, "as per the norm for typical production-centric companies, our organization tends to pool similar skill mix in a single separate group, in order to optimize the operations. PDK validation is a specialty that we have pooled in Fred's group, so that they can hone their skills and maximize their efficiency, etcetera. His group is staffed up to meet the average demand of the entire company – including the engineering and the ATI needs. The problem is that we do not have sufficient resources – particularly of a specialty skill mix like PDK development – to accommodate the inevitable *peaks* in demand. And when those situations arise, engineering development always gets de-prioritized relative to product work. This is understandable, and usually the consequence is that the schedule of the affected engineering work slips to accommodate the gap. So, on average it all works – sort of. Except in a case like Poopó, where we do not have the degree of freedom to slip the schedule. It is a real shame that we are missing this learning opportunity"

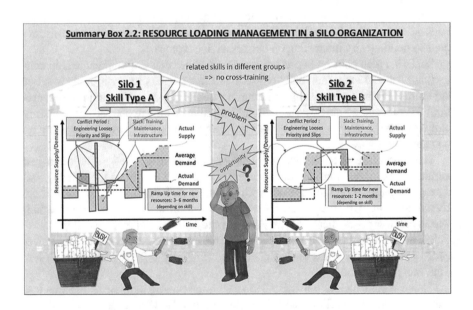

"Yeah, the problem is that getting the updated PDK was top priority for ATI – but not for Fred," adds Dr Cz. "That is one of the typical issues when working on things that are outside the mainstream – like our ATI initiatives."

"I understand, but…" I respond, "it wasn't Brian's fault. He really tried."

"Incidentally," I continue after a few bites, "how come that we could not do some kind of a work around? My layout tech was also pulled off Poopó – to work on something else – but she showed me how to do some of the polygon pushing"... I then add as an aside, "I like the term polygon pushing," and continue... "Anyways, with her help, I ended up doing both the design and the layout. I am sure it took me much longer that it would have taken her, but for my relatively simple structures, we managed to get it done in time. I did have to pretty much live here for the last few weeks, but still.... how come we couldn't do something like that?"

"Yes, I heard about your heroic effort. Thank you Jasmine," says Dr Cz and continues, "the fact that Fred is a bit protective of his turf and wants to control all PDK activities doesn't help. He is closed to the idea of spreading the expertise, and when the crunch time comes, there is no one with the right skill to pick up the load – like you did for the layout people."

"That is italicize problem," says Steve, with emphasis on 'the'... "In principle, the issue with an organizational structure that makes perfect sense in theory – like the matrix structure – is that it typically breaks down when you factor in some specialty skill mix. It just doesn't make sense to staple a whole PDK guy to every project. On the other hand, peanut-buttering the skill across a bunch of projects is hindered by ... err... territorial managers ... like Fred. We will need to address this somehow if we are to be successful with any of these Advanced Technology Initiatives."

"And it is not just PDK and Fred's group. It is all over the place," continues Steve, "for example, we had a similar issue with a DFM PathFinding study that we were trying to do. We needed some time from the library and physical design groups to assess the effect of 1D vs. 2D design rules (*TBB 2.3). With the way that the lithography technology is going, it is likely that the design rules for future technology nodes will drastically restrict all layouts to a single construct – disallowing bends and corners and other features which are normally used to compact a design. Implementing such 1D design rules would be very disruptive for design. So, I wanted to do a PathFinding study to quantify the impact of such constraints on chip area and to explore options for coping with them. It was to be an elegant study of implementing an IP block in 1D and 2D libraries. In that case we had the PDKs, but the design engineers assigned to the project were pulled off to deal with some product issue, and – poof – our PathFinding study is dead. So, we will be doomed to digest this kind of a disruptive change totally blind. Shame"

Technical Background Box 2.3: 1D vs. 2D Layout

The 'art' of layout design at the polygon level is to compact the design as much as possible and to use as few layers as possible, in order to minimize the Silicon area and cost. Traditional design rules allow constructs such as corners, dog bones, bends, different line/space widths, and orthogonal features within a layer – sometimes collectively referred to as 'two-dimensional' (2D)

(continued)

Technical Background Box 2.3 (continued)

layout – which were normally leveraged to compact the design. However continued scaling led to sub-wavelength lithography, double or quadruple patterning, and eventually compelled use of FinFET device architecture – none of which are compatible with these complex layout constructs. Hence one-dimensional (1D) layout rules – sometimes referred to as 'restricted design rules (RDR)' – were necessitated. The diagram below illustrates the typical layout constructs, allowed under the two paradigms (NB: for 1D rules = single direction, single width/space, no bends, no corners per layer...)

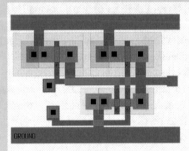 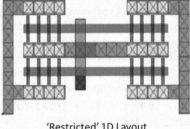

'Traditional' 2D Layout   'Restricted' 1D Layout

Note: material in the gray boxes is intended for those who are interested in more semiconductor technology and/or industry background information- and may be skipped by those who are not.

And with that, he gets up to go. I and Dr Cz both take an extra slice of pizza for an afternoon snack and wander back to our offices.

It is true – for the last few weeks I have been working 12+ hour days plus the weekends. I had to design the structures that were assigned to me, and on top of that, I had to learn the layout tools and practices so that all the required work would be completed on schedule. Two jobs – more or less – and one me. It was not the fault of the layout lady – she was pulled off Poopó – but she was nice and taught me what I needed to know and checked all my work. And I was willing to put in the extra hours, and she had enough time to do the hand-holding for me. On top of all that, Dr Cz asked me to help out in the lab with some measurements that needed to be done pretty damn quick – which also cost a few unplanned days.

But now all that it is done. I do feel good that I got all my structures on the Poopó chip. "Ha! You slayed it, girl," I say to myself, focusing on the kudos that I got from Brian and the team, rather than on the fact that everything else in my life has been pretty much on hold. I guess, I really did not have anything else going on – other than just chilling with friends. But I am feeling gross since I have not worked out in weeks, or gone for my usual morning run, and have meanwhile been surviving on health foods like pizza, burgers, and ramen. And my apartment is a mess, with much of my stuff still in boxes. Need to do an Ikea run. And sleep. I need lots of sleep…

Chapter 12
Managing Compensation (in a Matrix Organization)

"So, did you get your performance review," asks Panchali, one day, during our usual afternoon coffee. It's a few months later, and life has been pretty much a routine. I have managed to finally move into my place, and everything is out of the boxes and more or less in the right place. I am pretty happy in my new home. But other than that, a few uneventful dates and visits with mom and dad, work has been pretty much my life. And work has been good. Poopó chip just came back from the foundry, and I am now busily testing my structures in the lab – when I can get the time on the prober. As usual, the demands on the lab equipment exceed the capacity, and product-related tasks take priority. Lately, I have been working the graveyard shift to get some time on the prober. I don't mind it, though. I am excited to get the results.

"Yes. Seems like I will keep my job for a while more," I say, joking. "You?"

"Yes, I just got mine, and I am pissed," responds Panchali. She must be upset – seeing that she never swears. I have not heard her use even the mild expletives like 'damn' and certainly not 'pissed.' Something to do with the upbringing of proper young ladies in India, I suspect.

"It is just not fair," she continues. "The rumor in the Indian tribe is that the average bonus this time around is almost twice what I got. It is just not fair. I am going to talk to Cz about this," she fumes.

'Indian Tribe' is the usual tag for the engineers from India – many of whom tend to hang together. This as compared to the 'Chinese Clan' or 'Persian Horde' – the other large ethnic groups of RoCo engineers that tend to congregate and speak their lingo. I am sure that these terms must violate some PC rules – but everyone seems to be using them.

So, at the tail end of the next group meeting – which Dr Cz holds every couple of weeks to go over the active projects and to update us on the events in the company – Panchali says:

"Could you explain why our bonuses are reduced this time?"

"Huh? What?" responds Dr Cz. He seems surprised and taken aback.

© Springer International Publishing AG, part of Springer Nature 2019
R. Radojcic, *Managing More-than-Moore Integration Technology Development*,
https://doi.org/10.1007/978-3-319-92701-5_12

"Well, it seems that the average bonus given to the delivery teams are significantly higher," says Panchali, in a matter-of-fact way.

"Hmm. That cannot be. Let me explain how the system works, and then let's talk about it," responds Dr Cz, and continues, "as you know there are three components of the overall compensation package for engineers – the base pay, the bonus, and the stocks. The increase to the base pay – i.e., the merit increase – is driven by the performance rating and the position of one's salary relative to his peers in the industry. There is a formula that spits out the dollar increase that we get, based on our performance rating. The bonus is driven by the value of one's contributions over the evaluation period and, of course, how well the company is doing and hence how large is the pool of bonus dollars. And the refresh stocks that we get are driven by the long-term potential a given individual is deemed to have. All these components are expected to fit a normal distribution and have to fall within an overall budget defined by the company. Typically, there are some statistical deviations that can be negotiated at my level – since we are a small group – but the VPs are expected to fit a budget and a distribution. The way that the bonus is managed is that every VP is given a pool of bonus dollars based on a percentage of his overall payroll. Typically, a VP then apportions his pool down to his reports, and so on. There is a formula for this, too. So, by definition, the bonus pool given to our VP is the same percentage of his payroll as for any other VP. So, I just cannot see how our group can have lower bonus percentages than any other group."

"Well, be that as it may, I see that my peers in product delivery groups are getting better bonuses. And I think that this is not fair. We work just as hard as they do, so...?" says Panchali defiantly.

"Hmm. Let me go and check with Steve – but I just don't see it. Don't let me forget to get back to you on this," responds Dr Cz.

"Oh, I won't. You can be sure of that," says Panchali – still defiant. This is so out of character for her – I am thinking – she must be really upset.

So, at the next group meeting, Dr Cz opens with "Guys, I have some bad news for some of us – it seems that Panchali was right. Some of us got bigger bonuses than others."

"See!" says Panchali, sitting up.

"But" continues Dr Cz, before Panchali could say anything else, "it is not the way you think it is. It is not like the bonus pool for the delivery organization is larger than the bonus pool for the infrastructure organization. The issue is within the infrastructure organization itself. The bottom line is that the management felt that the impact of the people working on product-related deliverables – things for project Blu this time around – is higher than that of people working on the ATI projects. So, the bonus rating for those of us who have worked a lot for Advanced Technology Initiatives has not been as high as for those of us working on product deliverables."

"That is not fair," exclaims Panchali. "I work on things that you tell me to, and now it seems that when this involves ATI projects, I get paid less. Not fair!"

"Well, Panchali, I am sorry to say – but I agree with you," responds Dr Cz. "Steve tells me that he tried very hard. There is a leveling meeting with the whole VP staff that reviews all the ratings and makes sure that things are fair across the organizations. It was apparently hard to argue that a guy who produced learning which may, or may not, be important a year or two down the road is more deserving of a big bonus than a guy who has produced something that will drive our revenue a quarter or two from now. And with a fixed pool of bonus dollars, it was a zero-sum game – one guy getting more means that somebody else gets less."

Sullen silence in the room…

"Hmm," offers Dave, "that's a bit like the usual firefighting syndrome in all organizations. The guy who fixes some known problem – a fire fighter – gets the glory, and no one ever remembers the guy who has avoided a problem. That is just how things are."

"Makes you think that you are better off keeping some bug in your back pocket, until it becomes a problem, so that you can get all the glory – and a big bonus," adds Doug.

"Unfortunately – true… And hopefully not as bad as that, Doug. Presumably, we engineers do have some ethical standards. Not to mention all the design reviews that we have in order to bring our collective wisdom to bear," responds Dr Cz, sounding a bit defensive. "Anyways, I understand the management decision – in the sense that the impact on the corporate bottom line of product-related activities *is* higher than that of ATI activities. Therefore, according to the principles of bonus allocation, these contributions *are* more deserving of higher bonus."

And after a pause, he adds "but the intent of the Advanced Technology Initiatives is strategic – and people working on those should not be penalized. Our management clearly has not figured out how to weigh the value of strategic advanced technology work vs. product delivery work. Higher levels of management typically do value strategic work, but this clearly is not trickling down to the lower levels of the organization, and hence is not reflected fairly in the bonus allocation."

"So, what am I supposed to do?" demands Panchali, "the next time you ask me to work on something that is related to ATI? Should I just decline the dubious honor?"

"Well… no. You should never decline a work request from your boss – especially when that is me," says Dr Cz with a mock serious tone, trying to lighten up the situation. "But seriously – this is clearly a problem. We must fix it. Right now, I do not know how – but it must be fixed if we are to pursue these Advanced Tech Initiatives. I talked with Steve and he too recognizes the problem. Give us a chance to address it. I do not have a good answer for you now – but trust me – we will try to fix it."

"Humph," responds Panchali. She is all geared up for a fight, but with Dr Cz agreeing with her, there is no one to fight with. She is frustrated.

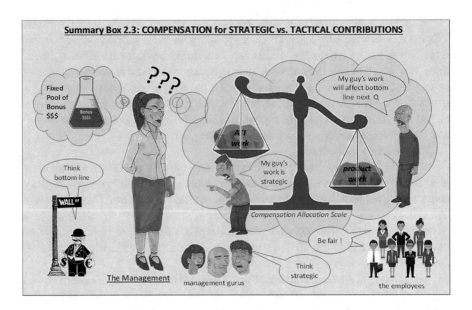

Summary Box 2.3: COMPENSATION for STRATEGIC vs. TACTICAL CONTRIBUTIONS

With this, the meeting breaks up, and we all go back to our offices. I close the door, and I am thinking to myself: "This really is not fair. Juggling Product Delivery vs. ATI assignments is hard enough as is. Product assignments often come as urgent requests, and you basically just have to drop everything in order to get it done. I – and let me guess, everybody else – prioritize product activities because they have to be done by a fixed date, the management wants it and is watching, and we all understand that the product show must go on. Which usually means that ATI tasks are stop-and-start kind of work that is done mostly after hours. I don't mind working the extra hours – after all, right now I have nothing else going on in my life. And I do enjoy the ATI work. But this really is unfair. I go and do all this extra work to get the ATI assignments done, and then I get penalized for my bonus dollars. WTF! Where does that make sense? It is sort of funny. I thought that people would be lining up and competing to get the advanced technology assignments. They are more interesting and surely, good for one's career. I was so happy when Dr Cz gave me the ATI structures for Poopó. Instead, it turns out that working on ATI is a handicap – at least in terms of corporate recognition and bonus dollars. How does that make sense?! So, like an idiot that I am – evidently – I work extra time on Advanced Technology stuff, and then I get to pay for it. I like work, but I also like having some time to myself. I like chilling in my apartment and binging on TV as much as anyone else… Or going to Big Bear. The squad went and spent a weekend up there a few weeks ago – and instead of going with them, I worked. Idiot… And Greg was there, too. Mmm… Greg. I like him. He is cute. I would like to get me a piece of that… Whoa JLo!! It is a good thing that there is no one to hear you! Listen to yourself. You sound like a real cougar. You are too young for that… I guess I really should do something about my social life… Never mind my social life – that is OK given my friends from the U – it is my sex life and romance that needs fixing. Too much work makes Jasmine a boring girl…" Hopefully all this was in my head rather than aloud. With that I turn off the light and go home. Feeling quite salty right now.

Chapter 13
Reorganization (into Line Organization)

"So, what's this offsite meeting that Steve called," I ask, sitting down with Panchali and Doug in the coffee room. Doug joining us happens sometimes – usually because he wants an update on the rumors. Panchali is our 'spy' – her term – in the Indian Tribe and the local wisdom is that the rumors from the Indian Tribe are usually the most up to date and accurate.

"Well," says Panchali, taking a sip of her tea, and drawing out her response, "Mmm…" She is enjoying this, I am thinking, because she has something juicy. "It is probably something to do with our acquisition of VisionX," she concludes.

"Yes, I heard that too," says Doug, "seems like they are basically a MEMS company with some cool proprietary technology for fingerprint sensors."

"So? RoCo buys VisionX – so what happens now?" I cut in. I understand from the various industry news sites and blogs that track the chip business, that takeovers are often followed by layoffs, and am therefore a bit concerned. I don't want to be canned. I have to cover the lease on my apartment, and certainly don't want to go back to living with mom and dad….

"Well," Panchali mumbles, "probably nothing. VisionX products do not compete with RoCo…".

"So? What's the rumor in the Tribe? C'mon, Panchali – tell," urges Doug.

"What? Who? Me? I know nothing. My lips are sealed," responds Panchali, playing coy. But then – after a pause – she adds, "I hear that the primary objective of the acquisition was to get their patents and other IP, rather than their product, and that the VisionX people will be simply absorbed into RoCo. They are a local company, so this can be done relatively easily, I suppose."

"Ah so so so," responds Doug "so Steve's meeting is probably to talk about that. But why do it off-site? And how come his whole group is not invited?"

"Really?" I ask.

"Yes, if you look at the e-mail list, it includes only some of his reports. And my highly scientific poll confirms this," Doug responds.

"Mmm," murmurs Panchali, suggesting that there is more she knows, "maybe some kind of a re-org?"

© Springer International Publishing AG, part of Springer Nature 2019
R. Radojcic, *Managing More-than-Moore Integration Technology Development*,
https://doi.org/10.1007/978-3-319-92701-5_13

So, by the time of the offsite meeting, a few days later, we are all a bit on edge – hyped up by the rumor mill. The fact the meeting is in a private conference room in Rancho Bernardo Inn – a fancy golf resort – doesn't help. It is unusual – but also exciting. Dr Cz and another manager in Steve's group, along with their teams are here. But there are also a number of other people who I have never seen before. Altogether, there is maybe 50 of us here.

"All right. Good morning, ladies and gentlemen. Let's begin," says Steve, standing in front, with the rest of us seated in chairs arranged in neat rows – cinema style. "I am pretty sure that you have heard all sorts of different things in the rumor mill, so I have brought us together to clarify things… RoCo has acquired VisionX – a local MEMS company based here, in Rancho Bernardo. That is no news by now…. What may be new to some of you is that as a part of that merger, VisionX engineering team will be combined with RoCo engineering." He then looks at the group of people I do not know, all congregated to one side of the room, and adds, "I hear that this will mess up the commutes for some of you guys. So sorry… But with the big money you all made in the sale of VisionX maybe you can get your chauffeurs to drive you while you read the morning paper and enjoy your Gray Poupon mustard sandwiches…" says Steve joking, trying to lighten up the atmosphere.

"Anyways," he continues, "as a part of that influx of new talent we have decided to re-organize the Technology Infrastructure Organization and set up a new Advanced Technology group – a team that is entirely dedicated to development activities." He pauses a bit, and then continues, "this group will report to me. And you all – the people in this room, are that team"… pause… "Congratulations"… pause … "I am sure that you notice new faces here. VisionX – meet RoCo, and RoCo – meet VisionX," he says waving in the respective directions of the two clumps of people in the room.

"So," he continues, "I thought that we should start by introducing ourselves to each other, and then let's talk about the new organization and our mission. That is why we are here… and to get a free RB Inn lunch, of course," he adds.

"I would like first to go around the room and for each one of you to say a bit about who you are and what you do. We will have a break and a lunch when we all can meet more informally, but let's begin this way... I will start."

And he then goes on to state his name and to summarize his bio. The rest of us follow suit, taking turns one at the time, standing up, and doing pretty much the same. Name… rank … commanding officer … bio. By the time this is done, it is coffee break time, and the two groups congregate, tentatively taking measure of each other. It is almost like a 1950's high school dance scene in the movies, with boys on one side of the room and girls on the other, talking and giggling among themselves, but wearily eying the other side – waiting for someone to cross the chasm and break the ice. Dr Cz takes the lead, walks over to the one guy from the other group that is gray haired, extends his hand, wearing his stupid grin. That seems to have done the trick, and pretty soon we all are mingling.

"Hello, my name is Atul Patel… and you are?" says this dude who walked over to where I was standing. He is a middle-aged Indian man, tinted glasses, slightly yellowing teeth, nice middle-age paunch, balding with a shiny pate, giving him a bit of a greasy look. And a very deep voice – on the account of which, I tag him 'Darth Vader' in my mind. We chat about what he did at VisionX – some packaging technology things … and what I do at RoCo. I hate to jump to premature and unfounded conclusions, but somehow, he gives me the creeps. Maybe it is his shit-eating grin – I think a more formal word is lascivious. The way he is leering at me makes me feel like he is checking me out. Maybe I am just imagining it? Maybe my tagging him 'Darth' has biased me….

"OK," says Steve loudly after a while, "let's continue." I am relieved to get away from Darth and return to my seat. "I would now like to outline where we have been and my vision for the future, and then let's talk about it," Steve says and puts up this slide:

Existing ATI: What went Right

- **Advanced Technology Initiative (ATI)**
  - Blessing from Management (CEO & down)
  - Budget (in principle) ~20% of Infrastructure Org
- **Advanced Technology Steering Committee**
  - Management input and guidance
  - Visibility: for and from VP Level
- **ATI Projects to date (DfM, DfV, DfTh)**
  - Approved funding
  - Approved Staffing

RoCo Inc

He re-caps the RoCo Advanced Technology Initiative highpoints and the way things work now – all presumably for the benefit of the VisionX guys. He emphasizes that the management – CEO level and down – sees the strategic need for investing in technology engineering with a horizon that is beyond that of product deployment. He outlines the role of Advanced Technology Steering Committee and points out that this is a mechanism used to ensure that the ATI effort is coordinated with other strategic directives, including marketing and business development, as well as engineering, roadmaps. And then he summarizes the intent of the projects that have been approved to date.

"That sounds very good," says Atul – aka Darth – who seems to be their lead, or at least their mouthpiece. "VisionX was much smaller and we didn't have anything formally structured like this. Any investment in future technology was managed on ad hoc bases. Mostly driven by individuals championing an exploration of this or that, and the management approving it on one-off bases… Sometimes…".

"Seems that Atul is my straight guy," Steve says jokingly and continues, "we thought that having some formal structure would be a good idea too. But some things did not go quite as expected." And he puts up this slide:

Existing ATI: What went Wrong

- **Execution**
  - o Hard to Compete for Resources vs Product Delivery
  - o Loss of Control over Development Projects
- **Motivation**
  - o Hard to Get the Recognition vs Product Delivery
  - o Loss of Share of Bonus $$
- **Focus**
  - o Hard to Compete for Attention vs Product Delivery
  - o Loss of Management Attention Span

RoCo Inc

He then summarizes the issues we have experienced, emphasizing that it is hard to manage Advanced Technology Initiatives within a matrix organization that is focused on product delivery, and illustrating the point with examples of our experiences. He adds another point – new to me – that Advanced Technology Initiatives naturally tend to move too slowly to keep the attention of higher level management, who are apparently more used to the pace of firefighting. So, it seems that the VP participation in the Advanced Technology Steering Committee meetings has been ebbing off.

Then, he says that the learning that we take from these experiences has to be baked into whatever we do going forward, and puts up this slide:

Existing ATI: Lessons Learned

- **Our Product = Learning**
  - o Test Chips – but need dedicated resources
  - o Pathfinding – but need skill mix span
- **Objective**
  - o It's all about <u>Integration</u> technology
  - o Partnerships across Supply Chain
- **Structure**
  - o Internal coordination (e.g. w/ marketing)
  - o Must be organized to manage the learning

RoCo Inc

And he says, "In my mind there are three critical lessons learned, so far. Firstly, our output is learning … Our primary tools are Test Chips and Pathfinding studies, and to be effective, we must have dedicated resources with a suitable skill mix." He pauses a bit and then continues, "Secondly, our opportunity and focus is integration technologies … This versus, say, pushing the envelope on any one given process module. So, we will collaborate with selected supply chain partners to develop innovative ways of combining mainstream technologies." He waits a bit for this to sink in and then adds, "And thirdly Advanced Technology effort must be coordinated with other parts of the company… Otherwise it becomes just another R&D ivory tower, whose output may or may not be used in a product."

"So, having said all that, our mission is," he continues and puts up this slide:

And he explains much of the thinking behind the original Advanced Technology Initiatives, putting the emphases on collaboration with the supply chain. He then outlines the key More-than-Moore type of target technologies that he wants us to focus on, highlighting the technologies for Integrated Passive Devices (IPD) and 2.5D and 3D Integration (*TBB 2.4). He then says that now that we have dedicated resources we have a certain capacity to learn, and using an analogy of filling a bucket with rocks and sand, defines these major projects as big rocks, but emphasizes that there is still room for the 'sand in our bucket of learning' that can be used to explore new and different technologies.

With that, he steps away from his laptop and asks, "What do you all think?"

Silence… We all stare at the slide and look at each other – maybe a bit confused. I look around the room and notice that Darth is staring at me. Hmm. I don't like it. Smarmy, I am thinking. He strikes me like a creep. I hope I won't be working with him….

"OK. Total agreement, I see," continues Steve. "I would like now to talk about where we go from here and specifically how do we structure the Advanced Technology group."

"You mean the organization?" asks Atul.

"Well, we have learned that matrixing Advanced Technology within a Product Delivery activities does not work. So, we now have the management approval to form a dedicated Advanced Technology line organization. We are a separate team with a separate mission that is not tied to any direct product delivery," Steve responds. "On the other hand, the classic challenge of such a line organization is maintaining the connections with all the other activities – avoiding isolation from the rest of the Company and becoming an irrelevant silo. This is a risk especially for activities that are not directly tied to product delivery – like ATI. Keeping those connections going will be one of our key challenges…".

"I get that," cuts in Dr Cz, "but now that we have this ATI line organization – how do we organize ourselves internally? With the addition of VisionX people to the existing RoCo team, we have quite a skill mix range. Packaging, Design, Test Chips, Characterization, Modeling… So how are we going to manage ourselves? Do we create a single pool of talent, and staff projects like in a matrix organization? Do we organize groups focused on specific target integration technologies – like mini line organizations? Do we do a bit of both?"

"Good question. This is what I think," Steve answers. "I believe that we should organize ourselves to align with our target customers. We – Advanced Technology – have customers like anyone else. So, the Integrated Passive Devices (IPD) technology is most likely to be used in the Analog/RF products. The 2.5D and 3D Integration technologies are more likely to be deployed by the digital products. So, I think that we should create two teams – one facing the RF product group focusing on IPDs, and one facing the digital product groups focusing on 2.5D and 3D Integration. And I think that we should resource these two teams so that they can operate more or less independently of each other. That way they can really concentrate on their specific mix of technologies."

Technical Background Box 2.4: Advanced More-than-Moore Integration Technologies

- *Integrated Passive Device* (IPD) is a type of a component which integrates passive elements (resistors, inductors, capacitors) into a single package, resulting in smaller, and often better, implementation of functions such as impedance matching circuits, harmonic filters, couplers, baluns, and power combiner/divider circuits. IPDs are especially of interest for applications where form factor, thickness, reliability, and power-performance are key constraints – such as for example some medical or mobile devices. IPD technology leverages aspects of advanced Silicon and/or Packaging tech-

(continued)

Technical Background Box 2.4 (continued)

nologies to build elements such as high-density trench capacitors, MIM capacitors, resistors, and/or high-Q inductors, usually on Silicon or Glass substrates. IPDs are an obvious case where value comes from superior integration technologies – drawn from different sectors of the industry – rather than through simple scaling of discrete passive devices. Example IPD chips:

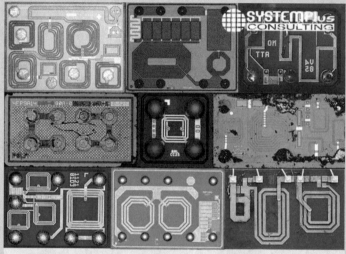

http://www.systemplus.fr/

- *2.5D Integration* – an integration technology where several Silicon dies are placed side by side in a single package. 2.5D technology is also typically differentiated by enabling a wider parallel interface between the die than would otherwise be possible, thereby producing power-performance benefits, in addition to form factor. This integration technology has been especially of interest for applications that require very high bandwidth memory (placing high bandwidth memory in the same package with a SoC processor), or use very large Silicon die (breaking the large die into multiple smaller dies and integrating them in a package). 2.5D technology leverages some aspects of advanced Silicon and Packaging technologies, and often (but not always) uses a 'Silicon Interposer' as a substrate that carries the high-density wiring required to interconnect the multiple dies. Conceptual 2.5D IC implementation is illustrated in the sketch below

(continued)

Technical Background Box 2.4 (continued)

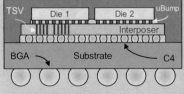
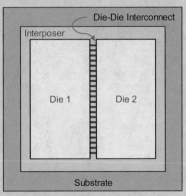

Differentiating Features:
* Multiple (2) die
* micro-Bump Die Attach
* Silicon interposer w/ TSV

From: 'More-than-Moore 2.5D and 3D SiP Integration', by Radojcic Riko, 2017

* *3D Integration* – an integration technology where several Silicon dies are stacked on top of each other in a single package. 3D technology enables very high-density integration and supports an even wider parallel interface between the die, thereby producing incremental power-performance as well as form factor benefits. This integration technology has been especially of interest for applications that require a very high-density implementation (e.g., stacking multiple memory dies), or tight coupling between dissimilar die (e.g., stacking sensor-on-logic or memory-on-logic). 3D technology leverages some aspects of advanced Silicon (e.g., Through-Silicon-Via) and Packaging (e.g., micro-Bump) technologies, and as such is another case where value comes from superior integration technologies, rather than through simple scaling of Silicon devices. Conceptual 3D IC implementation is illustrated in the sketch below

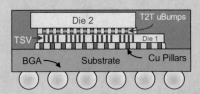
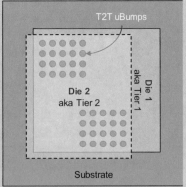

Differentiating Features
* Multiple (2) die,
* Tier-Tier micro-Bumps
* Through-Silicon-Vias (TSV)

From: 'More-than-Moore 2.5D and 3D SiP Integration', by Radojcic Riko, 2017

Note: Material in the gray boxes is intended for those who are interested in more semiconductor technology and/or industry background information- and may be skipped by those who are not.

"Hmm. That makes sense. 2.5D and 3D Integration do have some features in common, so it makes sense to put them in one group. IPD is different, so it makes sense that this is a separate group. And you are thinking that each of these groups should have the ability to design their Test Chips and do their PathFinding studies independently of each other?" asks Dr Cz, thinking aloud.

"But," Atul cuts in "firstly, do we have sufficient resources to fully staff these two technology development teams? And secondly, what do we do in case one of the groups needs some specialty that is resident in the other group, or neither group has a particular skill that is required?"

"Then," responds Steve, "we manage it at my level. We should certainly be able to slosh talent between the two groups. If and when necessary, we may be able to borrow special resources from external organizations. Why don't we work up the staffing plan for the two groups and see the gaps and the overlaps? I believe that we will find that there are gaps in the basic skill mix and limitations in capacity. So, in the short run we will have to approach the management for additional staffing. In the long run, perhaps we can think about some kind of a rotation program, where people from other groups rotate in and out of ATI on ongoing bases. Suitably selecting the incoming people would give us a way to continuously tune the skill mix of the team. And, the outgoing people would act as carriers who infect the rest of the organization with the ATI learning. What do you all think?"

"I like it," says Dr Cz, "the beauty of aligning with the customer organizations is that it should maximize the communications with them – as opposed to you being the sole interface to the outside world. Higher bandwidth and multi-channel communication…".

"Yes. That is my thinking too," responds Steve, "that link to the potential users of our learning – our customers – is critical. So, we might as well organize ourselves to facilitate it."

"That does makes sense," comments Atul tentatively, "we obviously do not know how RoCo works, so…. The next thing is to fill in the names in the boxes. Who gets to manage the two groups, and who is assigned where? Some of us have been working together at VisionX for a while, and it may be good to keep us together." Aha! I am thinking, Atul is angling to keep his VisionX team together and to be the manager.

Steve looks across the room and says, "Before we address the specifics – are there any other thoughts on the overall structure? What do you all think?"

Silence… Everybody is just looking at Steve, trying to visualize the new organization – and probably trying to figure out what it may mean for them individually. Steve has clearly thought about this for a while. For me, personally, all this organizational structure stuff is mumbo-jumbo. I just want to work on neat technologies. And I want to remain in Dr Cz's group – I am comfortable with him.

"OK. I will take that as full agreement, too," says Steve. "One more thing. I think we should call ourselves Advanced Technology *Integration*. This puts the emphases on the Integration – which is what we are all about. Our focus right now is on 2.5D, 3D, and IPD – all More-than-Moore integration technologies. And, as a side benefit, we get to keep ATI as our initials – since most of the outside management is used to it by now."

"Oh Excellent," jokes Dr Cz, "that way we do not even need to re-do our business cards."

"OK," continues Steve, "Why don't I take the action to work individually with you all, and I will come back with a specific organizational chart? But in principle, I am assuming that we are all on board with the idea of two groups within ATI, aligned to their respective clients and focusing on specific technologies."

And with that, he closes the meeting, and we all trickle into the adjacent room – where lunch is served. I try to sit with Dr Cz, who is at a table including a mix of new and old faces.

After salad is served, I ask him, "So, does this mean that we are to be re-organized and I may get a new boss?"

"Probably not yet," he responds. "Even though it would be good for you to try out some new managers, it is likely that you will be stuck with me for a while more."

"But, didn't Steve just say that he still needs to define the positions and postings," I ask.

"Ah, grasshopper…" responds Dr Cz. I like it when he calls me 'grasshopper'. I was confused by it, to begin with, but someone told me that it comes from this 1970's karate TV show. So, after I watched some of it – cheesy, to be sure – I got it. And I like it. It makes me feel like he is assuming a role of my 'sansei' – teacher – and is looking out for me.

"Yes, he did. But, an experienced manager like Steve does not go and open a can of worms like organizational structure unless he has done the rounds ahead of time. So, let me guess, he already has a pretty good idea who will be in which group, and has talked to people that will be affected, or at least to their managers. If you haven't heard anything, and I haven't heard anything – then things will probably be unchanged for you. I think Atul and some of the VisionX people will be the IPD group, and the rest of us will be the 2.5D/3D group. There are probably a few loose ends that still need to be tied up before Steve puts up a finalized Org Chart."

"OK, cool. I hope I am not one of those loose ends. I would like to stay in your group," I say and decide to keep my concerns about Darth to myself. I don't want to go to the dark side…."

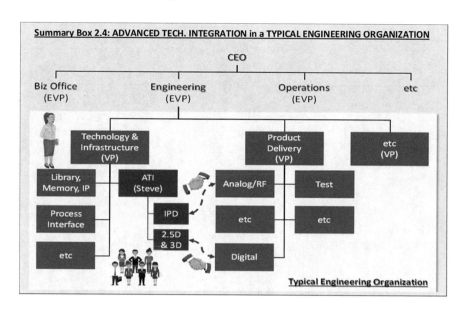

Chapter 14
The Challenge (*How* to Do the Integration Technology Development)

"So – now that we have a defined mission and have been given the resources – what do you all think are our main challenges?" asks Dr Cz.

We are in his group meeting. The new organization has been defined, and there are two teams in ATI – one focused on IPDs under Atul and one under Dr Cz. His group is now 20-odd people, including some of the VisionX engineers, his old team, and a few new additions. We are the so-called Advanced Technology SiP Integration team – SiP being 'System-in-Package' – and are supposed to be focusing on the 2.5D and 3D integration technologies. Dr Cz is excited – pacing around the room, gesticulating and waving his arms, making silly faces, and is generally all keyed up. Almost manic. But I like it – he is funny.

"Selling a new technology concept to a risk averse product team?" suggests Dave.

"Getting the supply chain companies to collaborate?" asks Doug.

"Finding the technology bugs? proposes Panchali.

© Springer International Publishing AG, part of Springer Nature 2019
R. Radojcic, *Managing More-than-Moore Integration Technology Development*,
https://doi.org/10.1007/978-3-319-92701-5_14

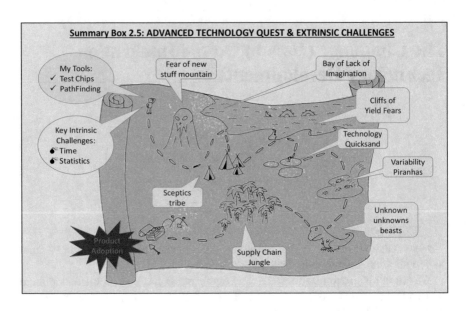

"Yeah, yeah, yeah," responds Dr Cz. "All true and things that we will have to manage as we go along. But all those are *extrinsic* challenges, so to speak. I mean what do you guys think are our key *intrinsic* challenges?"

Silence – he seems to have confused all of us.

"Well let's think about it," he continues, "our main vehicles for learning are the test chips. Right? Test chips are the source of data," and he writes on the board 'DATA' to emphasize the point. "Let's think about that. Test chips – if designed right – are very good for characterizing specific isolated device characteristics. But they are terrible at characterizing anything that requires statistics – for example, process variability – since we never run test chips in sufficient volumes. Similarly, test chips are not good at identifying the various possible interactions – unless there are specific structures explicitly designed to characterize those," he says and writes on the board:

Test Chips = DATA	
GOOD	*BAD*
Visibility into Discrete Characteristics	*Interactions – lack of integrated structures*
Testable and Observable	*Variability – lack of statistics*

"So, are you saying that basically we will need a product to fully de-risk a technology, including fundamental attributes such as variability and interactions?" asks Doug.

"And yield," adds Dave. "We need statistically significant samples to estimate yields – and that means costs."

"Yes," responds Dr Cz. "I think so. There is definitely a chicken-and-egg challenge here. Product teams may be reluctant to adopt a technology without variability and yield data – they would see it as high risk. But it is difficult to estimate variability and yield without statistics, and it is difficult to get the statistics without a product. You need to run many wafer lots – preferably 10's – to begin to see the tails of a statistical process distribution. Too expensive to run that many lots of test chips. Economic reality is that we won't get the statistics until there is a product... Chicken-and-egg. That, almost by definition, is our key *intrinsic* challenge... Especially if we are the first – or the only – user of a given technology option, so that we do not get to piggyback on some other source of data. And he writes:

Test Chip Challenge # 1: Statistics for Process Variability/Interactions/Yield Characterization

---

"What else, do you guys think?" he asks. And after a pause continues, "I think that Challenge #0 is time." He pauses a bit to let that sink in, and then elaborates "the length of a cycle of learning in our business. That is a problem even before we get to the lack of statistics," and writes:

Test Chip Challenge # 0: Duration of a Cycle of Learning

"Think about it," he says, "fab cycle time for modern process technologies is 10 to 15 weeks. Let's say we are rich and can buy super-hot-lot status and assume 10+ weeks. Since our mission is to explore chip package integration, we must bump and dice the wafers and assemble parts into packages – typically, 4 to 6 weeks. Test and characterization are 4 to 8 weeks – longer for new things, shorter for things we have seen before. We are good – so assume 4 weeks. And implementing a corrective action is typically of the order of 4 to 8 weeks – longer if it involves a design change. Sum it up and the critical path adds up to a cycle of learning of the order of 20+ weeks – at best. Tack onto that the time it takes to design a test chip in the first place – and given that we are really really good – this is another 8 weeks at best, plus something like 4 weeks to make the masks. So, we end up with less than 2 cycles of learning per year – even with the insanely optimistic assumptions. Realistically – even if we wave our hands and make a bunch of optimistic assumptions – it is more like a single cycle of learning per year! One! Uno!"

"That is depressing," comments Panchali, "but true, I guess."

"Well, on the positive side, this is an estimate for the first cycle, and subsequent turns may be quicker. You may not need a new test chip for every turn, and corrective actions may be faster. And obviously this assumes that whatever we do will involve something new in the baseline Silicon technology... There is a rule of thumb that it takes about 10 cycles of learning to go from start to full technology qualification. So, even optimistically, it will take us more than 5 years to take a technology from start to full qual. If we assume that we intersect the product prior to full qual – and given that we are really really smart, and my estimates are too pessimistic – let's half that. That says that it will take us 2.5 years to get a technology to a point where we can start talking to a product team. That is longer than life of a normal CMOS technology node. Meaning that if we target Silicon technology node n, by the time we have something, the product teams will be moving to node n+1, and possibly even n+2. Meaning that we are out of date before we start," concludes Dr Cz. "Modern day version of Zeno's paradox," he adds.

He lets this hang in the air, and then asks "So? What do we do?"

"Well," responds Brian – who, I am happy to say, is now also a part of Dr Cz's group and leading our test chip design team – "we will clearly have to use short loops rather than full flow test chip runs."

I have heard this term before. 'Short loops' are a kind of test chips that do not go through the full Silicon manufacturing flow but are instead special vehicles focused on characterizing just a specific feature or a few modules in the process flow.

"And/or," Brian continues, "seeing that the long pole is the fab cycle time, maybe we can anticipate the spectrum of features and variables we are likely to explore in Silicon technology and build a kind of a bank of full flow test chips? Build an inventory of suitable test structures – so that we do not have to go through the fab every time we try something different at the package level?"

"Excellent!" says Dr Cz, leaping to the board and jotting things down:

Test Chips and Opportunities for Acceleration of Cycles of Learning	
Solution 0: full flow test chip	Baseline source of data – run as hot lot
Solution 1: short loops	Vehicles that look at only a subset of features or layers
Solution 2: test chip bank	An inventory of Si test chips including variants (reflective of future Si tech.)
Solution 3: specialized vehicles	An inventory of technology-independent test vehicles

"What do you mean by that Solution 3?" asks Brian.

"It is a subset of your test chip bank idea. We can design variants of selected test structures in the latest technology, build extra lots of test chips, and bank them – presumably to represent what we expect to see in some future Silicon node. That's your Solution 2. But, some things may not require the latest Silicon technology at all – say thermal characterization or assessment of mechanical stress effects. We could design test vehicles that are entirely independent of the latest Silicon process node, build and bank them totally independently of a specific project. That is Solution 3."

"Got it," says Brian, as Dr Cz completes his table on the whiteboard.

"Excellent. So, clearly, we will need to think out of the box to develop the right kind of test chips for our integration project. We need something in addition to the conventional technology characterization test chips that we are used to – like Titicaca or Poopó. Variants of selected features. Short loops. Maybe specialized test chips to look at the package technology. Maybe specialized test vehicles to look at things like thermal and stress phenomena. And clearly, we need to make sure that these test chips are packagable, testable, and observable and all that. This may be hard – but that is why we have assembled an A-team like you all. So, no problem," he says joking. "We just need to think out of the box," Dr Cz concludes, and then follows up with "Agreed? Makes Sense?" to make sure everyone is on board.

We all nod and make positive noises – although I don't think anyone is sure *how* do we do all this. Clearly, thinking through the whole program top down demonstrates that there is a lot more to test chips than meets the eye. A separate art form, it seems.

"However you turn it," Dr Cz continues, holding up two fingers indicating that he is moving on to a next point, "it is unlikely that we will have the luxury of having all the data that we need. Reality of industrial life that we all live and love is that the window of opportunity for impacting a product-adoption decision rarely waits until all the data is available."

"Wait," I speak up. "Wait. Are you saying product decisions are made without data? This is so contrary to what I was taught. I am surprised that you have not been struck down by the engineering gods. Surely, good engineers make decisions based on data. That is like rule #1. What *are* you saying?" I ask, perplexed by where he seems to be going. I guess I am comfortable here and am not shy anymore about speaking up in meetings.

"Yes, grasshopper. Good engineers do make decisions based on data. True. So good engineers do everything possible to make sure that they have the data. And typically, on that spectrum between R&D and Product, the closer you are to the product end, the more, and the better, is the data. But what do good engineers do when there is not enough data? And let's face it – with this Advanced Technology enterprise, it is unlikely that we will have all the data that we would want when we need it. Process variability and statistics, reliability, and life test data, etc... We will not have that. So, in principle, we must be prepared for a situation where the data for a new technology that we are championing is incomplete at the time when the product guys are open to inputs… Think about it. Product development cycle is of the order of a couple years, and product ramp is another year or so. So, in principle, we need to be championing a technology some 2 to 3 years before it has been proven in volume manufacturing – which is when the data gets to be really good," he says and draws this time line chart on the board:

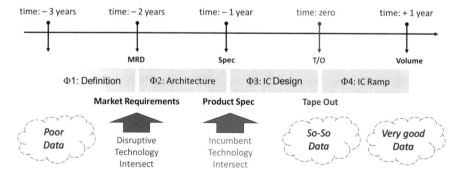

"These are the typical development phases for a modern SoC product, with the typical duration of about a year or so for each phase. With our kind of technologies, it is likely that we will need to tweak product architecture – so as to take full advantage of the features our disruptive integration technologies offer. So, we need to intercept phase 2. That means that the product guys will be open to new technology options around 2 years before the tape out. After that, it is too late – because the architecture is set to favor incumbent technology option. OK, say I am wrong – which I never am," he smirks, "and say that each phase takes half as long. It still means that the window of opportunity for adopting a new disruptive integration technology is a whole year before tape out. And typically, tape out cannot happen before technology qualification – which is when you get just the so-so data. So – that magic window of opportunity to intersect a product is open at best a whole year before qual, and more like 1.5 to 3 years before we have good statistical data."

"That is the ugly truth," says Brian, and all the others nod.

"So, what do good engineers do when they do not have all the data?" Dr Cz repeats.

"Pray a lot?" jokes Dave.

"No!... We guess... Or, to be more accurate, we backfill the gaps in data with our judgment. Technicians grind to a halt, robots go into keyboard lock mode, but engineers – good engineers – use their judgment!" Dr Cz says, getting quite passionate. "That is why we have engineers."

"Wait," I say again, "this has got to be wrong. We guess?" I am incredulous. This is so contrary to everything that I thought engineering was supposed to be.

"OK. Calm down Jasmine," he says, "yes, we do exercise our judgments, but – being good engineers – we base those judgments on modeling and simulations, rather than just picking it out of midair."

"So, good engineers backfill the gaps in data with models and simulations? OK – that I can buy," I say.

"Let's call that PathFinding. So, now that even Jasmine has bought off on the concept..." and then as an aside he adds, "aah, the young and the innocent"... and continues, "what do we think are the key challenges with PathFinding?" he asks.

"Well, we are going to need the right kind of simulation tools? Some may not exist yet – especially for multi-physics phenomena. And let's face it, reality is multi-physical," suggests Dave.

"Yes! True. This is clearly another thing that we will have to work on," responds Dr Cz, "but let's assume for the moment that we are not only good but also lucky. Let's assume that I have a magic wand and – poof – we have all the simulators available on the planet. This may not be all that silly – since EDA guys can be induced to work with us to make the right kind of tools. Now what?"

"OK. I like your wand. If we have the tools..." responds Dave, "next, we would need the models. Models that describe the technologies, the features, and the phenomena that we are looking at."

"Aha! Yes!" Dr Cz gesticulates and points at Dave "you da man!... The problem with modeling and simulations is the ol' garbage-in-garbage-out syndrome. If we do not have the models that represent the technology we are looking at, then, we are in effect doing what Jasmine just rebelled against – guessing. And even a very good educated guess, at best, would result in simulations that output only some kind of relative information. Sort of like doing thermal simulations only to conclusively demonstrate that when you cram more power into a chip, it gets hotter. Duh."

"But," retorts Dave, "sometimes relative information *is* valuable. It can guide us and give us insights in product sensitivity to given technology variables… You know – increasing parameter A in the process decreases product characteristics X, or something…"

"True," responds Dr Cz, "very true. Relative information is valuable to accelerate process-to-product tuning. No doubt. But in general – in absence of concrete data – in order to get an insight that is credible and actionable," he emphasizes 'actionable,' "PathFinding simulations have to be based on models that are calibrated and tuned to represent a given target technology."

"I agree," adds Brian, "my experience is that if these PathFinding-kind of simulations predict something that makes sense, then we all shrug and say that we knew that anyways. Or else, if the simulations predict something that we did not expect – then we write it off and say it must be garbage because the model is not calibrated. Either way – we typically need something credibly quantitative in order to act and do anything concrete."

"Yup. I like simulations and PathFinding as much as the next guy – but to be actionable they must be based on models, and models must be calibrated. They don't have to be accurate to the nth degree – but they need to be in the ballpark that is representative of the technology or the feature we are looking at…" Dr Cz concludes. "We need *fidelity* but not accuracy."

And then he follows up with a question: "And how do we calibrate the models? … aha!! Back to those pesky test chips," he says and slaps his forehead to emphasize the 'duh' aspect.

"So so so … then the data derived from our test chips is more for calibration of models for PathFinding than it is for demonstration of a technology?" asks Doug.

"Yes! You da man! I think that is a key thing to appreciate – and something that we have to bake into our test chip strategy. Test chips are there to calibrate the models, and models are there to enable the simulations, and simulations are there to drive the PathFinding, and PathFinding is there to make decisions... And when all that is said and done, we – hopefully – close the loop by running a separate test chip to validate our conclusions." He goes through this sequence, a point at a time, to really drive it home. We all nod, since it makes sense. Interesting – this top-down way of looking at things, I thought.

"In addition," Dr Cz continues, holding up three fingers indicating that he is moving on to a next point, "in my humble opinion, simulations are only as credible as the circuit that we are simulating. If you use a wrong type of a circuit as a simulation vehicle, the result you produce is judged to be irrelevant. So, in addition to test chips which we need to calibrate the models, we also need a number of representative Figure-of-Merit (FoM) circuits to use in our simulations. Things like suitable analog or digital blocks, memories, etcetera… right?"

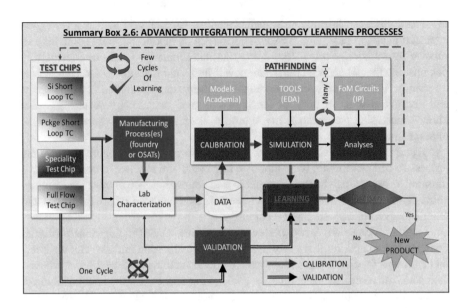

Summary Box 2.6: ADVANCED INTEGRATION TECHNOLOGY LEARNING PROCESSES

"Yes... that makes sense," says Dave haltingly, waiting to see where is Dr Cz going with this.

"Is that kind of analogous to testing of a new car engine? You need a good sporty car to demonstrate a high-performance engine, and you use a tractor to demonstrate a power diesel engine. If you use a tractor for a high-performance engine demo – you get laughed at," I ask, trying to cement the point in my mind. Everyone chortles – but in a good way. I am liking this group-thinking stuff.

"Very good Jasmine... But!" continues Dr Cz, "we have another chicken-and-egg syndrome here. It is hard to get the design enablement to design these figure of merit circuits in a technology that we just started playing with. And to use the PathFinding simulations the way that we want – to backfill gaps in the data, or to extrapolate from something that is known – we need these FoM designs very early on in the target technology life cycle. Chicken-and-egg. Make sense?"

We all nod. This sounds like a showstopper issue – but Dr Cz doesn't seem to be stopping?

"So, in addition to needing to think out of the box about our test chips – our *source* of data – we need to think out of the box about design of figure of merit circuits – our way of *extrapolating* the data. These circuits are a kind of test vehicle for us – like Jasmin's concept cars – and we need to invent ways of doing quick and dirty FoM design... So that we can change them easily, or tune them for a new technology feature, or something... So, to do the PathFinding on timely bases, we need FoM circuits early in the technology development cycle." He concludes and makes another table on the board:

PATHFINDING Simulations = Backfilling and Extrapolating from DATA	
STRENGTHS	*NEEDS*
Quick and Cheap	*Calibrated Models – to be Actionable*
Relative Sensitivity Analyses	*Quick n Dirty Figure of Merit Circuit*

"That is not easy," says Gorko – another guy who joined the team. I think he is Romanian or Bulgarian or something. Has a funny accent and a sort of squeaky nasal voice. A bit rotund. But the rumor is that he is a good designer... "All design flows and tools are optimized to make an accurate design and to compact it to minimum area. And designers are conditioned to do an exact design that will work and yield in Si. We do not have the tools – or the mentality – to do something quick-n-dirty. Especially in some fuzzy technology-to-be. We just don't know how to do that." He seems somehow defiant – almost proud of these good design practices – like a priest preaching the true faith.

"True," responds Dr Cz, "but, what are we to do? We clearly need quick-and-dirty FoM design – or else we cannot do credible PathFinding on timely bases. And without the PathFinding simulations, we are back to square 1, sucking our thumbs and sitting on a pile of inadequate data. So, Gorko – you lucky man – since you are our designer, your primary assignments is to figure out a way of doing exactly that. Quick-n-Dirty FoM design. Somehow... But note that we don't need a circuit that will function in Silicon – we just need something that we can use as a simulation vehicle and that is sensitive to the given technology attributes. So, we are lucky to have Gorko help us with that...," he concludes, pointing at Gorko, who is leaning back, arms crossed, looking skeptical.

"In principle," Dr Cz concludes, "if we are to be successful in our mission – we need (a) innovative test chips and (b) innovative simulation vehicles... Agreed? What am I missing? Anyone got a better idea?"

Silence... Can't argue with what he says, but clearly, we are all concerned about *how* do we do all that.

"OK. Good" Dr Cz continues, "let us now look at the whole ATI experiment in terms of *relationships*. The strategy is to achieve all this learning through partnerships. Right? ... And in all these relationships our goal is to be the '*grand integrators*'..." He makes the air quotes and lowers his voice to emphasize the point about the 'grand integrators'... pause ... "tadaaa!" ... he makes the fanfare sound. "If you think about it, in ATI we have assembled a very unique team with a rarely broad skill mix. We have design, device modeling, process engineering, packaging, system architecture, and what have you – all contained within a single, relatively small group. We are 'the mutants' who can bridge the skills and disciplines normally contained in entirely different organizations, or even companies. So, our strength –and differentiation – is to leverage that breath and to be the integrators. We are uniquely positioned to see – and capture – the value proposition offered by the various integration schemes. That is the plan – and our secret sauce," and he draws on the whiteboard a table like this:

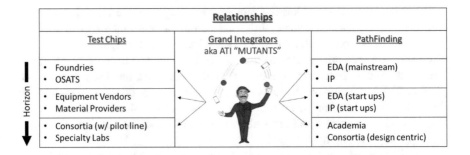

Relationships		
Test Chips	Grand Integrators aka ATI "MUTANTS"	PathFinding
• Foundries • OSATS		• EDA (mainstream) • IP
• Equipment Vendors • Material Providers		• EDA (start ups) • IP (start ups)
• Consortia (w/ pilot line) • Specialty Labs		• Academia • Consortia (design centric)

Horizon

"We have two principal tools – each with its own set of challenges – and we will use those to engage suitable partners in the supply chain. Test chips with the foundries and OSATs, PathFinding with EDA and IP guys… We are the grand integrators and will keep the global integration learning in house, while collaborating – and sharing – the local learning with each of our partners … There... What do you guys think?"

We all nod. The big picture strategy makes sense.

"Do we get ninja turtle costumes to go with that mutant tag?" asks Doug, smirking. Wow! Doug cracked a joke at work – amazing, I thought.

"No. That shade of green does not go with my coloring," jokes Dr Cz and continues "before I let you go, let's do one more thing. Let's think through what an overall project might look like. How do we do what we are supposed to do," and he draws a sort of schedule chart on the board that looks like this:

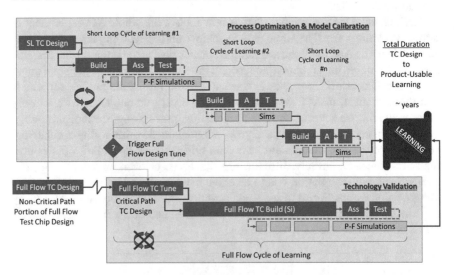

"It is all about holistic technology learning," he says, "and our challenge is to do it as fast as possible with the tools that we have… If we do it right, we will do most of the iterations and learning cycles using short loops and specialty test vehicles, in concert with the PathFinding simulations. And hopefully only one spin of the full flow test chips, in order to tie it all together and validate the conclusions we arrived at based on short loops and PathFinding. And then together with the product teams, we will work out how exactly we inject that learning into a real product and complete the technology and design de-risking required to fully qualify a product."

And with that he closes the meeting and we all file out.

"He does get worked up a bit, doesn't he?" says Brian quietly as we walk toward our offices.

"Umm," I respond, "The only thing that I got from that meeting was that we need to invent all kinds of test chips and figure out how to do quick and dirty FoM circuit design. But I have no idea how are we going to do it. You?"

"Yeah," he says "not easy. But it will be interesting."

Chapter 15
Test Chip Engagement (and the Assignment I Did Not Get)

"Jasmine has been a boring girl lately," I am thinking to myself. "Work, work and more work. Surely there has got to be more to adulting than this… At least I like the work, but still…."

I am sitting in my apartment, feeling lazy and in a retrospective mood. It is Saturday and happens to be that one day in a year when it actually does rain in San Diego – so I am enjoying being inside, cozy and warm, sipping my coffee, and staring out of the window. My apartment is finally done and looking the way I wanted it – and finally fully furnished now that I have got that cute sofa. And decorated – that coat of desert-sand paint on the one wall worked out – it goes well with my favorite mom-art. Looks good. Warm. And I like all the pillows that I have scattered about – to curl up with... The street outside is shiny and slick, deserted, and sort of tranquil. Even the usual dog walkers are nowhere to be seen.

"Hmm, maybe I should give up and get a dog?" I am pondering. "All my friends seem to be disappearing. Amy moved away. Dong-Wook is going back to Korea. Joni, Sam and Carlo are no fun nowadays – they are all lovey-dovey with the people they hooked up with, and rarely want to come out and play with me. Hanging with friends from high school and the LA neighborhood is… well… boring. Normally, once we get through the usual reminiscences of our antics in the past, there is not much that we can say to each other about the present or the future. A boyfriend would be nice… It is not that I need a man to define me – to borrow a phrase from the women self-empowerment propaganda – but it would be nice to have someone who is more available and focused on being with *me*. Someone to be close with and to share time and experiences. While in school I used to dream about travelling and seeing the world – and now that I have some money and could even get the time to do it – I have no one to go with. All that is a boyfriend's job – currently vacant… Things with Greg just went nowhere – so I gave up on him. He was yummy looking and fun – sort of diametrically opposite to Victor – not very cerebral. But, I mean – what does a girl have to do? I have done everything I could to show him that I was interested. Used all my 'bottom power' – to borrow Amy's African Pidgin term. But he just seemed to go hot and cold. One day he is fun and flirty, his eyes go all smoky

© Springer International Publishing AG, part of Springer Nature 2019
R. Radojcic, *Managing More-than-Moore Integration Technology Development*,
https://doi.org/10.1007/978-3-319-92701-5_15

and he dreams up some fun thing to do and asks me out; and then the next day he runs away and is cold and withdrawn. For a while I thought it was me – that I did or said something to turn him off. Then I thought that he may be just man-scared of commitments – which would be stupid, since I am not at a point in my life when I am looking for any commitment. Amy – who knows him the best – thought that he was into me, but that he is intimidated – it seems by my PhD and job and stuff. Go figure! That is so stupid and old school. It is rare nowadays, but – judging from movies and things – seems to have been common in my parent's generation. I just do not understand why a guy would be put off by my brains and degrees. Really? Why would a guy feel any less of a man if he is with someone who happens to be more educated, or makes more money, than he does? Makes no sense. Should be the other way around. And what am I supposed to do? Pretend to be a dumb sales-girl, or something? Really? And, you know, if a guy is scared off by that – then I don't want him. No matter how good looking he may be, or how much I am drawn to him. So, screw it. No Greg for JLo! No boyfriend. No friends who want to play... So maybe I should go and get a dog or a cat... But, I am not quite ready to be one of those proverbial crazy cat ladies... Not yet, at least...".

"Oh well…. Come on girl – time to get a move on. Go and work out, and then, lunch, and after that, read some of those papers on Through-Silicon-Vias… See… Jasmine *is* a boring girl."

Technical Background Box 2.5: Through-Silicon-Via (TSV) Technology

Through-silicon via (TSV) is a technology feature that forms an electrical connection between the frontside and backside of a Silicon chip – thereby enabling interconnection of die stacked on top of each other. The feature is formed by etching a very deep via (typically with depth-to-width ratio of ~10:1) in a Si substrate, lining it with an insulator and then filling it with copper (or sometimes tungsten). Then the wafer is thinned to expose TSV bottom, and a suitable end-cap is deposited for connection to on-die interconnect and/or a micro-bump.

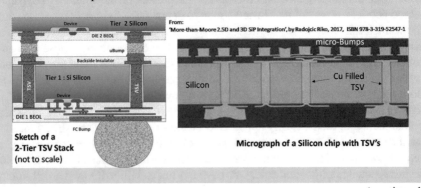

Sketch of a 2-Tier TSV Stack (not to scale)

Micrograph of a Silicon chip with TSV's

(continued)

> **Technical Background Box 2.5** (continued)
>
> TSVs can be defined in a Silicon wafer along with just a few interconnect layers to form a 'Silicon interposer' which is used for 2.5D integration. Alternatively, TSVs can be inserted in an IC to enable stacking of active die used for 3D integration. Typically, TSVs are 3 μm to 10 μm in diameter, with the Silicon wafer thickness of 50 μm to 100 μm, respectively (thicker wafers with bigger TSVs are typically used in Si interposers. 3D stacking typically requires small vias and thin wafers).
>
> *Note: Material in the gray boxes is intended for those who are interested in more semiconductor technology and/or industry background information – and may be skipped by those who are not.*

Over the last year or so, I have been digging into those TSVs (*TBB 2.5). Seems like a promising technology for chip-package integration. When I started I thought, 'what's the big deal: etch some holes in Silicon wafers and fill them with copper?' But seems like there is a lot more to it than that. So, through the year, I have read a zillion papers on TSVs; I have designed a bunch of TSV test structures, characterized many of them in the lab, and defined some electrical models and am now looking at their thermal and mechanical behavior.

So, it was not a surprise when a while later I was invited to participate in a meeting with SanCo – a Japanese company that has been working on a version of the TSV technology. I have read a number of papers that they published – an interesting twist on the typical TSV process flow... These meetings with companies from all over are not unusual, and I guess a part of our strategy – and a benefit of being a big company. I like meetings with external companies – not only because we get cookies, coffee, and a catered lunch but also because they are a great way to learn. The visiting company typically presents their technology, and the discussion afterwards is an excellent opportunity to get an insight into what they are thinking and where they are going. Sometimes they even leak what some of our competitors are doing. Normally, in these meetings I am mostly a fly-on-the-wall, so to speak, since it is usually Dr Cz who presents our slides. But I do get to ask questions and participate in the discussion – and to learn. The visiting team typically consists of a local account manager, a few of their technologists flown in from the home country, and some senior corporate manager. You can always spot the account managers – they are the guys with power ties and a professional handshake, who, following the introductions, withdraw into the various latest electronic gismos that they always seem to have. And the technology dudes are the guys who do the presenting – sometimes seems like reciting from memory. I guess they may be going through the material for the nth time. Or maybe they are just jet lagged. They tend to get animated when we get into the technical discussion. And the corporate manager tends to look fierce – quietly sitting and listening, wearing a stern expression, and occasionally grunting their approval – typically until lunchtime. That is when the discussion turns to collaboration opportunities and the projects that we could do together. I guess this 'dance' is a part of building relationships with the supply chain partners, which Steve goes on about. It's cool.

So SanCo presented their TSV technology – interesting – and we had a very good technical discussion afterwards. Lunchtime discussion focused on the possibility of demonstrating their technology using our test chips, in order to characterize their unique TSV and to explore the opportunities for integrating it with our baseline Silicon and Packaging technologies. Great meeting.

After the meeting – when the visitors are gone – Dr Cz comes to my office and asks, "So, what did you think?"

"Unique TSV process technology... Interesting – if it really does work," I respond.

"Yeah, I thought so too," he says. "With their approach the TSV does not have to be done in the foundry and could be inserted by an OSAT. That would give us new degrees of freedom for integration. And it feels to me like this process is intrinsically cheaper than the standard approach. I like it... The problem may be that SanCo is not one of our regular suppliers...".

"So?" I ask, "can't we just add them to the list of suppliers?"

"Oh, that is a huge deal. The operations guys have a boat-load of procedures and requirements that a company must meet in order to be qualified as a RoCo supplier. It is unlikely that SanCo would want to go through all that without a business deal in sight – or that they would meet all of our requirements," he responds.

"So – in that case, why are we talking with them? Sounds like we cannot use their technology – no matter how good or bad it may be," I ask.

"Good question... Well," he explains, "if the technology has some tangible advantages – in terms of cost or performance or what have you – and gets baked into some product of ours, then the pains of certifying a new supplier would be worth it – for them and us. Or, alternatively some kind of a deal could be worked out where SanCo licenses their technology to one of our existing suppliers. Or – since multisourcing is a requirement for our ops guys – some mix of the two."

"Multisourcing?" I ask.

"Yes, RoCo sourcing requirement is that there must be at least two manufacturers for every key technology. This of course tends to keep the price down, and also mitigates any sourcing risks – in case one of the suppliers has a problem of any kind."

"I see. So, they are pitching their technology to us – I mean to ATI – as a form of a long-term sales effort? I suppose that from their perspective we are a sort of a back door into RoCo products?" I ask, always wanting to get a better understanding of the business and relationship side of things.

"Yes. Well... a back door to RoCo, and a way of selling their technology to the general market. Typically, companies like SanCo partner with us for two reasons. One – they get to demonstrate their technology using our test chips and our integration and characterization know-how. Oftentimes specialized companies like SanCo do not have the breath like we do and may have hard time getting the right kind of test chips or the right kind of wafers and stuff. So, we bring technical value to a deal. And two – they get to use our name in their collateral. RoCo is an important player in the industry, and using our name gives their technology credibility and is therefore of value to them, even if none of our products end up using the technology, per se," responds Dr Cz.

"Collateral? You mean like brochures and things?" I ask.

"Sort of. More like White Papers. If we do a project together, they like to use the collaboration as a way of advertising their technology brand, so they publish a so called White Paper that advertises the results, and leverages a mention of RoCo as a contributor."

"I thought you said that RoCo legal and marketing guys don't like it when we publicly endorse a supplier. Your guideline was that if we publish our work we must concentrate on problems and challenges, rather than solutions. We do not publish results and data, you said?" I ask.

"Well, yes. Business guys know that there is value in RoCo public endorsement – so they like to trade this for something else. Like a lower price or something. But for an engineering engagement at a pre-product technical level, like what ATI does, we can authorize our partners to publish these White Papers – since the engagement does not involve products or product design. This is one of the things that we trade for access to their technology, and oftentimes for the significant free engineering effort that they have to put into these engagements," he responds.

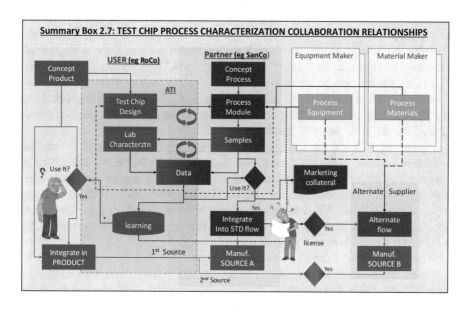

"I like SanCo technology," he continues. "I think we should engage with them and do an in-depth characterization. If it really does work – it would be a good thing for us. I think I'll go and talk to Steve and start the contracting process."

Then, a few weeks later, Dr Cz sends an email setting a meeting in his office. This is a bit unusual – he normally just drops by if he wants something. So, I am a bit anxious, and on the day of the appointment, I go over and knock on his door saying "Hello Dr Cz. So, am I in trouble?" – only partially joking.

"Good grief Jasmine – no. Wait – why? What have you done?" he jokes back and continues, "I have good news and bad news."

"Oh?" I ask, sitting down on the chair he waved at.

"The good news is that we are engaging with SanCo! Turns out that the legal paperwork – the NDA and a Master Contract, and what have you – already exist from some prior engagement with another group at RoCo. So, all we have to do is a new Statement of Work, and we are done," he says.

"That is excellent. So?" I ask, waiting for more.

"And," he continues, "I talked with Yoda-san, their CEO, and it turns out that they are also eager to do a joint project – and they want to do it, like, asap... We checked, and it looks like our test wafers – with the structures you designed – can go straight into their line. So, we agreed that rather than sending samples back and forth across the ocean, the best thing is that one of us goes there and works with their team in their pilot line and lab for a while. That way we can avoid a lot of paperwork, and accelerate all the cycles of learning, and quickly tune the process for best results."

"Cool!" I respond, now getting a bit excited. It would be really so awesome if I was going to do this. In Japan! I would love going to Japan! Wow. Real savage. Fingers crossed.... And an image of me standing in front of one of those typical pagodas, surrounded by cherry trees in full glorious blossom, was already forming in my mind.

"Yes, we thought so too. Excellent learning opportunity and the best way to collaborate," he says and continues, "but, the bad news is," he halts and stammers a bit, "that we cannot send you," he concludes.

Bummer! I am disappointed and am feeling a bubble forming in my throat and ask, "Why? I have been working on TSVs a while now, and I designed and characterized the test structures that you want to use?"

"True," says Dr Cz. He looks away and fidgets a bit, scratching his beard, and recrosses his legs, as always, perched on his desk, and adds, "errr... look, there is no good way of saying this, so I am going to be direct. This assignment is an excellent opportunity to build a real in-depth relationship with SanCo. If the technology is all we hope it is, that relationship will be critical. So, the assignment really involves a lot more than doing the lab work – it is more like being an ambassador who will form and cement that relationship."

"Yes," I say. "So? I can be an ambassador too."

"Well," he continues, "yes, you probably can. But – look – over the years I have been to Japan many times, in hundreds of meetings, and the only time that there was a woman in the room was to serve the coffee."

"So? Are you saying that I don't get this assignment because I am a woman?" Now I am getting angry.

"Errr. I guess... maybe," he stammers.

"That is so unfair," I cut in. "I mean – I cannot believe you are saying this," I say, getting more angry, as it sinks in.

"It may be," he responds defensively, "but, look – my job is not to fix the world. My job is to get this project done the best way I know how. Things may have changed since the last time I was there, but building relationships in Japan involves more than office hours. And since they are a top-down kind of a society where position is correlated to seniority, it typically requires dealing with *old* men. In my judgment, I am sorry to say, you – being young *and* a woman – would be a handicap."

"That is so unfair. You know that I am the right person for this assignment. I know the technology and the test chips better than anyone here. And I can drink beer or sake and be as polite and correct as anyone else," I inject, not sure if I am raising my voice or not.

"Yes, you probably can," he responds, lowering his voice trying to calm me down. "Technically, you are a slam-dunk choice. But this is different. Building that relationship, like I said, is part of our key objective. Relationships – especially personal relationships – are instrumental for doing business in Asia. And that requires something else. It is not anything tawdry or dirty, but I believe that they – especially the senior guys amongst them – would just not be comfortable with a young woman. It is just the way it is."

"That is so very wrong. And unfair. I have been breaking butt here and now you pull this assignment from me because I am a woman?" I say, now being not only angry but feeling hurt – and so disappointed. Shook.

"Look Jasmine, I know … and understand," he says trying to sound sympathetic. "You are the right technical choice – which is why I am talking with you. And I do feel awful about this. But like I said, my job is not to fix the world but to get the project done. Normally I would go myself, but I cannot get away for six to eight weeks that this would take. So… I need your help."

"What?" I say, glumly – trying to swallow my anger and hurt.

"I need you to work with Brian and teach him some of the TSV technology, your test structures and stuff. Help us get him ready for this assignment," he says and continues, "I know that this is like adding insult to injury – but it is the right thing for RoCo. Brian is the right choice for this assignment. And – trust me – we will make it up to you. There will be other assignments. Please, just trust me," he says, almost pleading. He is looking so earnest and sad – looks like *he* is about to cry.

I am busy tamping down my anger and disappointment. I nod, get up, and leave – feeling like slamming the door behind me. I need to be alone – hope Elvis is not in our office. But he is – so I pack up and go home early. I don't think I have ever left work before 6:00 pm – and here I am, 3:00 and at home. I am pacing around my apartment and thinking – fantasizing – how good it would feel to storm into Dr Cz's office and tell him that I quit. They don't want to give me this assignment because I am a woman – so screw them all!

The next day I go in and look for Panchali – I need her advice. I decided last night to do nothing rash, so… She is her usual cheerful self but, after taking a look at me, asks "what's up Jasmine?" with a concerned look on her face. So, I explain the situation and ask her if she would quit if she was in my shoes. She pats my shoulder and says, "Oh dear, that is terrible. I am so sorry that this is happening…" We talk about it, and I am feeling like she is really sympathizing and sharing in my disappointment – and anger. In the end she says, "I have learned that when considering big changes, it is best to go *to* something, rather than *away* from something.

The satisfaction you would feel by quitting would last for a short time, but you have a job for a long time. So, if I was you, I would absolutely quit – but only if and when I had a better job. And you need to figure out what is a better job than the one you have."

I am cooling off a bit now. It is nice to have a sympathetic listener. I mull over things, thinking about how I like living here, how I enjoy advanced technology and working with the team, and am remembering some of the good things. I would hate to leave all that. But I am still angry and hurt. So, I ask her what she would think are my opportunities of transferring out of Cz's group – a way of keeping the good and venting the bad.

She thinks about this and says, "You mean to Atul's group? Oh dear. I doubt that you would find more sympathy and support there," and she goes on to explain that the rumor in the Indian Tribe is that Atul has a history. "They say that in VisionX he had a reputation for a wondering eye, and maybe even wondering hands. I don't know the details, but from what I hear, he is not a good boss for a woman."

So, it seems that my instincts about Darth were correct "Really?" I ask. "He does make me feel uncomfortable – the way he leers at me. You know how you get that creepy sense sometimes? How come someone doesn't do something?"

"Eh, Jasmine, you are so young" she says. "Even if it wanted to, the company cannot fire or discipline a man just for making someone feel uncomfortable. And even if it could, people just do not report it. Are you going to go and talk to HR?"

"No, no, there is nothing that I can report. Nothing happened. It's just how he ogles me," I say. But I am feeling … well … trapped and somehow helpless. Very frustrating.

"You know, in a perverse way that is a good thing. Here most – unfortunately not all – of these kinds of things stop short of behaviors that are actually actionable. Back in India – it is horrible what men get away with. You don't want to know," responds Panchali, shaking her head.

Then, after a while, she says, "This is what I would advise you to do… Take a few days off. Go somewhere by yourself and think through things. There is no such thing as a perfect place. Everywhere there are some pluses and some minuses. RoCo ATI and Dr Cz's group has some plusses – but now this is happening to you and you are upset, there are also the minuses. Atul ogling you is clearly a big minus, but maybe there is a position there that is so good that it makes up for it? Not to mention that there are other jobs – inside and outside RoCo. You have to figure out for yourself what kind of compromises you are willing to make. Find the sweet spot with tradeoffs that you can be happy with. And then go for it full speed, and make it happen."

And that is exactly what I did….

Chapter 16
PathFinding Engagement
(and the Assignment That I Did Get)

I went to Casa del Zorro in Borrego – a swanky desert resort in the back country of San Diego. Normally I would not feel good about spending the kind of money this costs me – but I thought that under the circumstances, spoiling myself may be OK. Besides I didn't really feel like going camping alone. And going with someone would defeat the purpose – since, for once, it is the alone me-time that I was looking for. So, I spent a couple of days there – me with a bunch of old retired people and a few moms with young kids. I guess, midweek in April, that is only to be expected. I hung by the pool; I went hiking; I had a hairdo, mani, and pedi; I ate and drank – probably too much. And I pondered stuff. A funny alignment of things, I thought. On one hand, I hear guys saying I am an attractive woman, but it seems that they would prefer it if I was less smart. And on the other hand, I have my boss saying I am smart, but it seems he would prefer it if I wasn't a woman. So, what the hell is a girl to do? Talk about a no-win for me. I am as aware of sexism as any girl – especially in engineering – but I did not expect it from someone like Dr Cz. Frustration and disappointment on all fronts. I allowed myself to wallow in, well, anger and self-pity and to vent the frustration. But after a while, I went into an analytical problem-solving mode.

The thing with Greg would normally not bug me. There are people like that – people I categorize in my 'write-off' bucket – and I have learned long time ago that it is best to just ignore them. It's not like I am going to convert them and make them into reasonable people. The reason that Greg's reaction bothered me is probably amplified by the fact that my friends are either dissipating or settling down into family life – and I am feeling left out. Clearly, I need to expand my social network beyond my college friends – meet new people and make new friends. The trouble there is that I have been hiding at work and really just haven't had the time to do stuff. I like people at work – but they are mostly in a different phase of life and do their own things. So, they are colleagues and not friends. According to the general pop wisdom – the thing for me to do now is to pursue my own interests *outside work* and to join in with other people who have similar interests. Maybe I should look at all those clubs and activities that crop up in various email broadcasts and ads – but

© Springer International Publishing AG, part of Springer Nature 2019
R. Radojcic, *Managing More-than-Moore Integration Technology Development*,
https://doi.org/10.1007/978-3-319-92701-5_16

that I always ignore. Ski trip to Mammoth Mountain? Kite surfing lessons in Mission Bay? Hiking up to Yosemite? Or maybe there are some kind of travel clubs or group tours? Someone was telling me about biking tours in Europe. That would be cool. I should look into all that; I want these things – but not alone, by myself.

And the thing that bugged me with Dr Cz and the assignment to SanCo is the sense of unfairness of it. Sure, I resent the sexism, but it is a sense of loss that hurt. For a minute there, I got my hopes up about going to Japan – only for it to be snatched away. Because I am a woman – of all the reasons. That *was* pouring salt on the wound. Not much that I can – or want to – do about being a girl. But on the other hand, I get Dr Cz. It is like me deciding that I am not going to fix all those 'write-off' guys. He has decided that he is not going to fix sexism in Japan. I can see that. Doesn't make me feel any better – but I can see it. And if Cz is right about the attitudes there – would I want the assignment? Probably not. So, changing jobs now does not make any sense – other than that the venting would feel good. Or at least, changing jobs for that reason does not make sense for me right now. On the other hand, looking at different jobs – to have an option and a choice – does make sense. I should look into that. Meanwhile, I shall carry on at RoCo and ATI and stinkin' Dr Cz. Maybe throttle down on the hours, and get more me-time.

And so, I did help Brian get ready for this sweet assignment in Japan. I felt bitter about it – but swallowed my disappointment and did the best that I could. It was the right thing to do. Besides, it is not like it is Brian's fault. He was cool about the whole thing. We agreed to set up a bi-weekly phone call, once he got there, so that I can stay connected to the project as it unfolds and help him from here.

So, I have been analyzing some of the data that they generated and am now building a model that describes the SanCo TSV characteristics. People have started talking about integrating SanCo TSV into our process flow, so we need the models to do the PathFinding studies. It is possible that SanCo TSV may have unique sensitivities – which would dictate different constraints, which in turn may have an impact on how they are designed and used. I really like doing that – a nice way of bridging all sorts of disciplines and considerations and getting a good, broad understanding. Interesting. So consequently, I have allowed myself to get sucked back into the work vortex – again – and am mostly back to working crazy hours. Back to being a boring girl.

So, one Monday – I remember it because it was a day after Joni's wedding - I was in the middle of one of those simulation runs, when Dr Cz knocks on my door and comes into my office. My head was hurting, and I was feeling salty. There really should be a rule against having weddings on Sunday afternoons. The morning after, the happy couple is fine and enjoying their honeymoon – but what about the rest of us? Nursing a headache *and* trying to work. The wedding was fun – except for that bozo that Joni picked. I just don't get what Joni sees in him. The good thing – scratch that – the great thing was that I met Mariano there. He is someone's cousin or something, Argentinian or Peruvian or from somewhere down there, and does something in the finance industry. He told me all that, but I wasn't paying attention – I was too busy enjoying his baritone voice, the way he softens his p's into b's and y's into j's, his smile, his manner, his Antonio Banderas-like looks, his laugh. And all that *before* I was buzzed. I liked him.

Anyways, Dr Cz comes into my office, sits in Elvis's chair, and eyes me with a kind of a smile on his face – like he knows some joke or something.

"So, grasshopper," he says. "I have good news and bad news. Which do you want first?"

"Oh no, not again. Are you going to pull me off this PathFinding stuff because I am a woman?" I ask, only half-jokingly. We have talked about the whole SanCo thing several times and now can almost joke about it – but carefully since both of us know that this is a sensitive subject.

"Yes, I am afraid that I am... that is the bad news," he says, trying to be serious.

Uh-oh. I am now trying to focus and ask, "and the good news?"

"How would you like to go to Switzerland and do it there?" he asks – a huge inane grin on his face.

"Wha...? Are you serious? Come on – don't play with me like that," I say, maybe somewhat phlegmatically – partially because of the last time and maybe because of my... errr ... hangover.

"Before you get too excited – consider it very carefully," he says. "This is serious – and big. As you know, we have been collaborating with the IDC Institute in Lausanne on their PathFinding methodology. In my opinion, it looks quite promising. I have talked with prof Doganis, and we agree that the way to take the methodology to the next level is to tighten our collaboration. We also need to get the EDA vendors more involved, and he feels that this too would require someone like RoCo to push them. So, he has opened a position in the institute for a visiting collaborator and hopes that RoCo would fill it. Interested?"

I am not sure that I am processing so I say "Maybe... Probably... Tell me more. When? How Long for? To do what? Why me?" still trying to focus.

"Well, let's talk about the position first and then about the assignment. And then, you need to think about it and then let me know what you decide. This is not the kind of a thing you do lightly," he says.

"The posting would be for about a year," he continues, "you would carry on as a regular employee of RoCo but your salary would be supplemented by a stipend from IDC Institute. That is to say – no sweet ex-pat package for you. Switzerland is one of the most expensive places in the world, and Lausanne is one of the most expensive places in Switzerland, so between your salary and the stipend you should be OK – but you would have to live modestly. If you can sub-lease your place here, it should be easier. So, this is really not about money, and you will probably feel like you are reverting back to being a poor student. But, on the plus side, you would have a good travel budget and would be expected to come back here for regular synchups on something like quarterly bases."

"That sounds ok," I say, getting more excited in spite of trying to keep cool.

"Well, hold on," he says, "truth-in-advertising, so to speak, compels me to tell you more about the down side – so you go into this open eyed. Typically, the problem with this kind of assignment is that you are absent – or more or less absent – for a whole year. That is a long time in our business. We – ATI – would of course want you back, but a year is a long time and things may change. We may get re-organized and there may not be an ATI in a year. The company may hit a rough spot and there may be some cost cutting. Or god knows what. This kind of a position tends to be vulnerable. So, in all honesty, I cannot guarantee you that your position – or even your job – will be waiting for you when you get back. In fact, in all honesty, if the company decides to watch its pennies, there is potentially some risk of being laid off while you are out there. This could be a one-way trip for you. That is not our intention – but there is some risk."

"Uh-oh. Well that does not sound good," I say.

"Think about it… Now let's talk about the mission," he says and continues, "As you know, PathFinding is all about using models and simulations to get a sense of the impact of given technology attributes on selected product characteristics. This prof Doganis has been working on a methodology for abstracting various aspects of a design, so that we can do this kind of simulations relatively easily. He seems to have developed a way of avoiding the need for a detailed design of Figure-of-Merit circuits. Remember – we talked about all that?"

"Yes, yes," I say "so, he has a way of avoiding the chicken-and-egg conundrum for design of FoM circuits in some future technology. I think I saw some of his papers a while back…".

"Right," Dr Cz responds. "We want to get a deep understanding of this methodology. And the thing that we really want is to implement it in a set of EDA tools, so that we could develop a complete PathFinding flow. Input some technology characteristic or constraint, turn the crank and – presto – out comes an estimate of key chip characteristics, like performance, or area. Would that not be cool, or what?"

"That would be awesome," I respond, now getting excited. This does sound like an interesting project.

"So, the position would require you to learn the methodology, work with prof Doganis to tweak it for our needs, and interface with EDA companies to capture it in a tool. We will also push the EDA guys from our end and encourage them to collaborate. And hopefully we end up with something we can use. If it all works, we will have a powerful methodology and a flow. Realistically, at this stage, we will probably end up with a few novel point tools and perhaps portions of a PathFinding flow – something that will require a lot of manual handling. Hopefully, it will add up to be something that we can use as a backbone for PathFinding studies that we talked about."

"That all sounds really interesting," I say, now beginning to feel like jumping up and down with anticipation. "Whoa, calm down girl," I remind myself – but at least the headache seems to be gone.

"Well again Jasmine – think about it," he continues, "as always there are risks and downsides. Pursuing this PathFinding methodology would take you out of the mainstream, so to speak. If the methodology takes off, you are golden. You would be like Moses bringing the tablets from the Mount. But if the methodology does not take off – you may be branded as the one working on that stupid PathFinding idea that went nowhere. It may be a black mark for your career – a sort of handicap that may take you a few years to overcome."

"Oh. Hmm. I understand," I say. Things are never simple.

"OK," he concludes, "think about it. Come and visit me if you need more info or want to discuss it. Take me out for a beer if you want to talk off the record, so to speak. But let me know in a few days. I want to pursue this, and if you don't take it, I have to find some other sucker... errr... volunteer for it," he says jokingly. "Actually – you are a very good fit Jasmine," he adds, "not just technically but also from personal point of view. Without an ex-pat package an assignment of this length is, realistically, unacceptable for people with families. Gorko, for example, could fit this project reasonably well, but his kids' commitments make it impossible for him. So – think about it. Once you give us a nod, it will take us a month or two to put everything in place: the contract, work visa, what have you. So, it is not like you will have to climb on the plane next week." And with that he leaves.

Wow! A roller coaster! Some heady highs and scary lows. Umm. I need a rock to sit on and ponder all this – and I especially need a clearer head. Switzerland! Pathfinding with Doganis. Awesome! But, back to being poor – since I don't think that I can sublease my place. Not so nice. And the risks? I don't want to be canned, I don't want to be sidelined in my career. A lot of variables. So, after some thought, I broke it up into two categories: professional and personal.

I first focused on the professional part and took a few weeks to get my head straight. I carved out time with Dr Cz and we had a few brainstorming sessions. We agreed that if prof Doganis's approach really does work, it would go a long way toward addressing the key concerns with the whole PathFinding methodology. We also talked about the fact that the easiest path to implementation of the PathFinding methodology would be to leverage the existing EDA tools – but that we would need to customize them for ease of use and speed at the expense of accuracy. The idea being to easily synthesize many candidate solutions to reflect different process or architecture constraints. We agreed that Dr Cz could take the lead with approaching the EDA companies and identifying a suitable partner and that I would focus on working with Doganis. My initial role would be to provide him with a target technology specs and an architectural concept, based on RoCo's products, and then to work with him on implementation of his methodology. And we would compare notes as we go along and hopefully expand the partnerships with EDA guys to nucleate an entire ecosystem for this PathFinding methodology. That all sounded really exciting to me. Dr Cz said that he thought that some of the EDA vendors had development centers in Europe – something to do with the European government and EU support – which may be especially interested in participating in this project. So, depending on how things evolve, I may need to do some shuttling around Europe to make it all work. And, he said that he thought that being a woman would in fact be an asset rather than a handicap, said something about everybody wanting to work with the pretty lady – especially if she is sporting a RoCo badge on her chest. So, all that sounded really positive and interesting. If PathFinding ends up going nowhere, I would have still learned a lot about EDA tools and chip design methodologies. So, I told him that I am interested, and to start the wheels rolling – but that I still need time to fully commit.

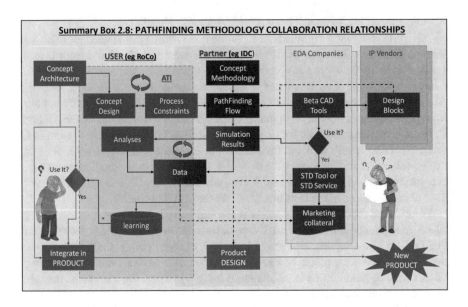

And for the personal part, over the last month or so, I talked to mom and dad, and Mariano, and some of my friends. Mom and dad were neutral – sad to have another of their kids stationed on the other side of the world but supportive of whatever I choose. Most of my friends said I should go – just for the thrill of it. And Mariano? Well, we have been spending a lot of time together since Joni's wedding – had a few introductory dates and, I confess, I waved my normal self-imposed rules about waiting, and we already have had a few … mmm… very nice nights together. Anyways, Mariano was really cool. He actually tried to understand what the whole thing was about, weighed all the pros and cons, and ended up saying "Djaz" – that is the pet name that he gave me in his funny pronunciation – "when all is said and done, the worst that can happen is that you will need to find a new job. But, finding a different job is something you will likely have to do anyways – sooner or later – in your career. So, in a bigger scheme of things – it is not a big deal. Nothing irreversible or permanent. And, if you do go, you will have a great time, learn a lot, and experience something entirely new. The value of that is permanent and irreversible. So – why would you not go?" And then he said that on the other hand he would miss me and that he would need to come and see me there – often – because, well, he said "because, you, Djaz, are special, and I intend to pursue you." He is funny with his old-world formality. I like that man.

Just my luck that this romance seems to be blossoming now – when I am thinking of leaving for a year. But, between my trips back to the home office and his promised visits, we may end up seeing each other every month or so. That is not too bad. And with Skype and things, maybe we can keep whatever we have going. And if it does fall apart, it probably was not meant to be.

So, I decided to do it. I am off to Lausanne, Switzerland!

Chapter 17
a Manager: Steve's Soliloquy (Circa + 3 Years)

I am a risk taker. In the modern corporate world, a safe career path is to do the conventional thing. As they say, 'no one ever got fired by buying IBM' – an adage derived from the time when Big Blue was the pinnacle of the industry and represented buyers' safe choice. But I'm a risk taker – I like being the guy who champions the risky options – even at the peril of getting dinged for promotions or missing out on other corporate reward mechanisms. Playing it safe is not my thing – and hopefully not the thing that the big guys always want. It's like a game of golf. Last weekend, I'm standing on the 3rd tee at Torrey Pines, needing to decide if I play it conservatively and aim off to the side of the green, or if I go straight at the flag, even though it is heavily fronted by a deep sand bunker and likely will cost me extra strokes if I 'hit the beach'. I can visualize different targets, think about how good I would feel if I make the aggressive shot, and assess my confidence in making the different shots and the impact if I miss the target. Then I decide which way to go and fully commit and focus on that target. Golf is a game with little negative consequence of failure, so it is easy for me to go for the high-risk, high-reward target. I probably make the high-risk shot less than 20% of the time, but one great shot sure makes a memorable day. So, of course, I decided to take the risky shot. I missed, but Michael Jordan said that he missed more than 9,000 shots in his career, he lost almost 300 games, and 26 times he was trusted to take the game winning shot and missed. He failed over and over and over again – but is still regarded as the greatest basketball player of all time. Great success does not come without risk and without repeated failures. No risk – no glory.

So, I tend to be a risk taker. I enjoy the thrill – as long as it doesn't hurt anyone else. Unfortunately, it seems that my daughter has picked up the risk-taking thrill from me, judging from the picture she just posted on her Facebook page – of her bungie jumping off a bridge somewhere in Africa. Maybe there are limits to my affinity for risk-taking, and some risks are not so palatable for me.

© Springer International Publishing AG, part of Springer Nature 2019

R. Radojcic, *Managing More-than-Moore Integration Technology Development*,

https://doi.org/10.1007/978-3-319-92701-5_17

Fundamentally career choices are risk-versus-reward decisions. From the beginning, the things that attracted me to the development side of semi-conductor technology were the novelty and the risks that go with that. In grad school, my professor – a world leader in the field of microwave semi-conductor devices – was constantly driving us to take technical risks and to try to set new records for performance by developing new materials and device structures. "You got to go crazy and try wild new things," he kept saying. His laboratory was intoxicating with innovation, risk-taking, and out-of-the box thinking, and I guess that spirit stuck with me throughout my career.

Another thing that I have learned about myself is that I get turned on by the excitement of learning experiences when working on 'the new stuff'. In the semiconductor industry in particular, where over the past 40+ years, a new technology node came along every 2–3 years; the learning is nonstop. Throughout my career, I found that continuous learning is a must, regardless of whether I was working on the latest and greatest technology, or squeezing the last bit of performance and cost out of a mature manufacturing process. But it is technology development that provided me the greatest opportunities for new learning. I worked on over a dozen MOS technology nodes, plus several 'More-than-Moore' technologies – including the RF compound semiconductors, 3D packaging, integrated passives, tandem solar cells, and magnetic memory. For all these projects, advanced technology development inevitably required the integration of many different disciplines – electronics, chemistry, material science, design, test, EDA, packaging, reliability, etc. I found that I enjoyed working with a strong cross-disciplinary team to solve difficult problems and find unique solutions that only a diverse team can realize. It is great to work with good engineers from varied disciplines and to appreciate a problem from many different angles.

So, I was very blessed throughout my career to get to work on advanced technology development. This proved to be a great way to satisfy my attraction to learning and risk-taking. For a few years, I tried working as a process engineer on a manufacturing production line. Not for me. Maybe it was those middle-of-the-night 3rd shift 'line-down' telephone calls that further turned me away from manufacturing and toward development.

Of course, engineering is about finding practical solutions to challenging problems in order to make money – not just setting records. I realized this soon after I joined my first company in Palo Alto. Even though that first job was supposed to be in a research lab, the realities of life in a revenue generating enterprise were a stark contrast to grad school. However, surprisingly, the breadth of new learning opportunities that I experienced there, as compared to the grad school, was staggering – mostly because there were so many experienced industry experts working in a team environment. I guess the revenue generated by a commercial entity is a necessary requirement to aggregate a pool of talent like that. So, I learned that commercial success is as much a result of, as a path to, the learning opportunities.

As I matured I have come to see myself more as a broad multidiscipline integration guy than a deep expert in any one area. This also seems to be well suited for technology development – and for group leadership. So, I turned to leadership and management fairly early in my career. I have the battle scars from all my experiences to remind me of the importance of balancing technical and management considerations in decision-making on complex tasks. I, of course, needed the techie skills to enable me to recognize patterns and to extrapolate, to anticipate consequences of options, to identify false information, to make fine distinctions between conflicting data, etc. But also, I needed the management skills to manage a team of people, to harmonize their work with the rest of the company, and to take the context and office politics into account. Plus a practical business acumen to create commercially successful technologies. These are important factors – even in highly technical world of semiconductors.

I joined RoCo a few years ago with the expectation that I would get an opportunity to champion advanced technology projects. When I first came on board, there was zero technology work on anything with a horizon that is more than 6–12 months ahead of product introduction. Essentially there was no advanced technology activity, no technology roadmaps, and no development phase methodology… nothing. What a great opportunity, I thought, since clearly a rapidly growing company like RoCo will have to implement some advanced technology engineering to survive and succeed as an ongoing entity in the future.

The initial challenge was deciding how to define an Advanced Technology Initiative, and how to sell it to the management in order to get their sponsorship and financial support. One of the trickiest aspects of this whole Advanced Technology endeavor is understanding and synchronizing with the expectations of the sponsoring senior managers. We live in a top-down world, and it is the CEO and the various key VPs who are the guys with purse strings. And, after all, it is their job to steer the company, and to select the various strategic directions – right or wrong. Of course, they need all the help that they can get. My job is to give them that help – by offering them risky, advanced technology options that hopefully, at least sometimes, fit their intentions.

In the early days of my career, R&D was conducted by academia – such as Stanford University – and by captive research labs, such as Bell Laboratories and IBM TJ Watson. Large corporations did in-house all their development and manufacturing, often constructing their own wafer processing equipment, developing their own materials, and coding their own design automation tools. Even semiconductor companies, such as Fairchild or Texas Instruments, followed this 'vertically integrated' approach to R&D. But this vertically integrated approach to R&D died off years ago. The cost and time of developing the next generation of semiconductor technology became too great for any one company to carry the whole load by itself. Instead, a model emerged where various stakeholder entities pooled together resources for precompetitive research. Beginning in the 1980s, various consortia, including Semiconductor Research Corporation, Microelectronics and Computer Technology Corporation, SEMATECH, and IMEC, were formed. The best of these still exist – and even thrive. So, an active membership in one of these consortia seemed like a valuable ingredient of any advanced technology endeavor that a company like RoCo would need. The learning derived from a membership in such a consortium would supplement any in-house proprietary IP development and could be further rounded out by relationships with suitable universities, to end up with a holistic R&D infrastructure. Insights derived from consortia and academia would provide the foresight and direction necessary to gauge the value and risks associated with any novel technologies and thus help identify the technologies that may be worth pursuing for productization in-house. So, I first needed to sell the management on the idea of joining an R&D consortium. At the time, this was quite a significant step – there was no precedent of Fabless entities such as RoCo participating in these consortia. But I worked at it. I organized presentations from candidate consortia to RoCo management, facilitated RoCo VPs' visits to consortia forums, set up a pilot project, and talked about it one-on-one with all key senior management, all to remove the barriers of unfamiliarity and to ensure the appreciation of the value that a membership would bring. It worked! Consortium membership runs into millions of dollars – but after a while, the management agreed to support it. We joined IMEC, sent our engineers as resident assignees there, and got to actively participate, along with the other members, in shaping the technology roadmap by steering the consortium activities. Check in box number 1.

Next, I was thinking, we need management support for in-house advanced technology efforts. But the kind of investment that is needed for this can be prohibitive – and scary for a company that is not used to investing in technology. Failure is an essential part of any engineering development effort – that is how you learn. But the management is typically conditioned to avoid failures. How much failure can they stomach? Should we go after the aggressive 'game-changing' technology, even though failure is likely? Or should we be more conservative and pursue lower-risk but less differentiating technology advances? What if different managers up my management chain have different opinions about how much risk to take? Do I just line up with the highest manager in the chain, or with my immediate boss? What percentage of the funded projects are expected to be successfully incorporated into products – 10%, 25%, or 100%? Does this expectation shift over time, and how much is it dependent on factors such as targeted market, current market share, or competitive threats? What is the maximum time horizon for new technology development – 1 year, 3 years, or 5 years? If the expectation is 100% success on high-risk, high-value, short-duration projects, with minimal project funding, then any Advanced Technology Initiative is doomed from the start. So, I thought that we need to develop a methodology for managing in-house development activities – and management's expectations for the results of such an effort. Some mechanism, with a set of checkpoints, that would gate the investment based on the value, status, and risks of a given development effort. Such a mechanism would then allow the company to gradually dribble bite-sized investments while down-selecting projects to meet the risk-benefit profile that is palatable to the management. I discovered that gauging that risk-benefit profile that is palatable to the collective management is, in fact, an art form – and that it was next to impossible to get an agreement on anything that is generic, abstract, or permanent. Not even for motherhood-and-apple-pie kind of endeavors. But I have found that the management collective could agree on supporting a mechanism that would allow it to reach consensus in the future – or at least a preponderance of opinions – on concrete technologies and specific investments. So, I proposed an Advanced Technology Steering Committee and a set of technology maturity metrics. And they supported it. We got ourselves a structured mechanism for managing the management – including their investments and expectations. Check in box number 2.

Next thing that we needed is to define a strategy for in-house technology development – how much technology intimacy is the management prepared to fund. There is a sort of a spectrum of choices here. On one end of this spectrum are the IDMs who do process development in their in-house fabs and pilot lines; on the other end are the plain vanilla Fabless entities who use off-the-shelf technologies offered by the foundries and OSATS. Where on this spectrum does RoCo management want to be – and how much is it prepared to pay for it? Investing in a fab or a pilot-line would not make much sense, I thought. This would just make us into a beginner IDM. It would hardly make sense for us to invest a billion dollars only to become a second-rate IDM. The strength of the IDM model is that having in-house fabs and proprietary process technologies gives them the freedom to create customized technology solutions which optimize their products. The risks of that model are that they may end up with a white elephant – an orphan technology that ends up being uncompetitive – and, of course, that they have to carry the burden of the cost of fabs and development engineering. On the other hand, the strength of the Fabless model is that it leverages the collective expertise of a supply chain and of course that the company is not burdened by the risks or the massive capital investments. The downside of the Fabless model, however, is that they typically do not have the degrees of freedom to customize the technologies – they have to use whatever the supply chain offers and typically cannot differentiate their products through manufacturing technology. So, is there something in-between, I wondered? There are fab-lite companies who have proprietary processes that are run in foundries or OSAT lines. Sometimes they even acquire private equipment which they consign at the foundry or OSAT, in order to support their proprietary technology. That is an in-between option, I thought. Are there other strategies that would be open to RoCo? Well, I thought – another advantage that RoCo has is that it is relatively large – a whale of a customer in the supply chain. Would it be possible to leverage that size – that buying power – to get some degree of freedom to optimize a technology? What if we partnered with a selected supply chain company and collaborated on developing some customized option that would be good for our product? Maybe we could pay them for the extra engineering effort? Maybe we could barter with them and get that extra engineering effort for free if we commit to use their technology? Maybe we could obtain some, or even most, of the value of customized technology, by creating a unique integration scheme? Or maybe some mix of all of the above? A large Fabless entity has access and power in the supply chain that is typically not afforded to our smaller competitors. It felt to me like there is an opportunity somewhere in there – an opportunity to create a custom technology by collaborating with supply chain partners to create proprietary integration solutions. Potentially by tweaking a bit their regular off-the-shelf processes? So, I decided to explore the options. I embarked on a round of worldwide visits with a number of foundries and OSATS. I also thought that using custom integration technologies would require custom design tools – you just cannot 'eat' a technology that you cannot design for. So, I included the EDA companies on my list of visits. Being old actually helped – because I had a personal network that I could use to set up the meetings with the right people. It meant that for a while there I was living on the airplanes, but the feedback that I got was extremely encouraging. Everybody that I met was not only open to collaboration but was in fact eager to establish such relationships with a large Fabless entity like RoCo. So, the next challenge was how to sell the idea to the management. Usually, it is best if power players in the management think that an idea is theirs, or at least that it is not a surprise. So, I thought that we would need to generate some buzz in the industry, which would 'soften the beaches' – to borrow a military phrase – and then do a full frontal attack. So, we presented a number of papers at the usual industry forums, participated in various discussions and panels, and talked to several reporters and bloggers who cover the business – in general we mounted a media blitz. In the end we succeeded to generate a buzz in the industry about 'Integrated Fabless Manufacturer' (IFM) model – large Fabless entities who have proprietary integration technologies. I also asked my contacts in various outside companies to talk to their counterparts at RoCo. And then I floated a few proposals through the Advanced Technology Steering Committee for specific integration technology projects that we could peruse. And it worked! We got the approval to proceed. Check in box number 3.

And now, I am concerned about managing the actual advanced technology effort – managing the people and the skills and executing on the projects. I am concerned about the mechanisms for rewarding people engaged in ATI activities. After all, realistically, final success of an advanced technology effort may not be realized until years down the road – when a technology is implemented in a product and is in volume production. Will anyone at that time remember the engineers who developed a technology option in the first place? And I am concerned about managing failure – an essential ingredient of any high-risk project. By definition, many advanced technology projects will not proceed to productization – either because of some underlying technical flaw, or because of product teams' natural reluctance to absorb the risk of adopting something new, or because the supply chain may lack adequate investment for cost-effective manufacturing, or any of the many other possible reasons. Would failure of an advanced technology project have an adverse impact on team members' performance ratings and career opportunities? Like for Jasmine, for example. Would leading a failed ATI project taint her reputation, regardless of the true underlying factors, and impact her career growth and opportunity in the company? The perception about the engineer's performance is highly dependent on the 'buzz' picked up from the office network. In response to a young reporter's question about how he expects history to remember him, Winston Churchill replied: "Young man I expect history to view me very favorably because that's the way I intend to write it!" But, it is the winners who get to write the history – not engineers who have worked on a project that failed due to no fault of theirs. So, I am concerned about the company's perceptions of advanced technology team. I know that one – maybe obvious – way to address these concerns would be to convert the advanced technology team into some kind of a separate R&D organization. R&D organizations are typically funded by some form of an internal flat tax imposed on the business units, and people are managed to entirely different criteria – criteria which are often closer to academic than industrial metrics. I thought about it, but, in order for the strategy to work – especially the aspect about collaboration with the supply chain partners – it is vital that we stay closely coupled with the RoCo buying power in the supply chain. That, and maintaining close links to our eventual customers, means that we must reside in the same organization as the product delivery teams, which in turn means that our people are inevitably compared to the people working on product delivery and directly affecting the revenues. So, this means that I must pay extra attention and focus on communications with the other groups in the delivery organizations. These guys are busy and often feel like they do not have the time for anything other than some product-related fire that they are dealing with – so I will have to work hard at it. I know that the key to increasing others' perception of the value and contribution of your team depends on the quality and frequency of your communication with them. But I also know that office politics, self-interests, organizational silos, communication blockades, quirks of personalities, empire-building ambitions, and tribal infighting can all factor into the tone of those communications.

So, I will need to practice all of my management skills and executive communication disciplines and be aware of politics in order to manage these perceptions. I will need to leverage my personal network, and any other means, to set an expectation that failing often, and failing fast, is part and parcel or any advanced technology effort. Understanding the power dynamics of the organization and respecting the agendas and turf of powerful managers will be essential to retaining management sponsorships. This is particularly important for advanced technology projects since management interest can wax and wane over the course of a multi-year project. Personally, I find these political games distasteful and distracting and wish that they were not so important. But that is a part of my job. I am committed to navigating these political complexities and ambiguities with integrity and resolve, and I accept the handicaps that come with my choices. I understand that a career in advanced technology is not the fastest path to corporate glories. Naturally, engineers working on the front line, closer to revenue generation, and where risks of failure are lower, and fire-fighting opportunities more frequent, are more likely to get promoted. I know. I have decided many years ago that I would rather work on new and interesting projects in advanced technology, even if it means the opportunity of promotion is lower. A key part of my reward is working with the team and continuously learning new things. And – on occasion – seeing our advanced technology projects make a positive difference for the company and maybe even the industry. When that happens – it is like that difficult golf stroke – it sure makes it all worth it. I will persevere.

Part III
I, Leader (Product Intersection and Handing Off the Learning)

This section, set some time later, finds Jasmine – now an experienced engineer – leading a project at RoCo semiconductor company. The Advanced Technology Integration team is struggling to demonstrate the value and risks of the new integration technologies and to transfer the learning accrued over the years to product delivery teams. Jasmine discovers that the dividends from the efforts and choices made in the past come in different forms and sometimes from unexpected corners.

Chapter 18
I'm Back!

"Welcome back Jasmine!"

"Well, thank you. Although I was never really gone. I must say – I missed you guys. I loved Europe, the food, the people, the lifestyle … and … errr… Dr Cz, of course the work. But I sure missed the team here... Even, you, the Double-D-Duo" I add teasing Dave and Doug with the somewhat edgy tag that someone ascribed to them – since they share the same double-barrel initials and their expertise overlaps, so that they do a lot of stuff together. "And the weather… And especially the IPA. Europe has some fine beers – but nothing as good as the local microbrew IPA we have here."

We are in Duffy's – officially celebrating my return – but I suspect, just an excuse to get together. Dr Cz's group does seem to have a European-like propensity for socializing off-site – must be the years that he spent over there. Funny how Southern California has the weather but no tradition of sitting outside – while Europe has the tradition of street cafes and no weather for it. Maybe it is because if Duffy's had outside seating, we would be enjoying a fine view of the parking lot, whereas in Europe it would be some piazza, or lake, or something.

"So, how is prof Doganis?" asks Dr Cz.

"The energizer bunny? That is the tag I gave him in my mind. He is great – goes at 100 miles an hour, never stops talking, laughs like a cartoon mad scientist, and keeps coming up with all sorts of novel ideas. Love the guy – but I can take him only in limited doses," I say. He – prof Doganis – is a sweet, short, round, balding Greek guy in his 60s but very hyper and always excited about some new project or others. Nonstop manic – hence his tag. I enjoyed working with him and his PathFinding methodology but found keeping up with him exhausting.

© Springer International Publishing AG, part of Springer Nature 2019

R. Radojcic, *Managing More-than-Moore Integration Technology Development*,

https://doi.org/10.1007/978-3-319-92701-5_18

"Oh, Dr Cz, let the girl catch her breath. She just came back and is probably jet lagged. Talk work on Monday…" cuts in Panchali, with a smile, and then turning to me "So Jasmine, much more importantly – how is Mariano?"

Everybody loves Mariano. Panchali met him only once and always keeps asking about him.

"He is good," I reply "He came to Lausanne a week ago – supposedly to help me pack – and after we stuffed all my worldly goods in a whole of two suitcases, we had a great time saying good-bye to our favorite places. We flew in together yesterday. All that packing was sort of silly, because half of the clothes that I brought back I will hopefully never wear."

"You flew in Yesterday? And you are at work today?" asks Doug, shaking his head.

"Yeah. Flights are cheaper mid-week. And I find that jet lag gets me on day 2 or 3, so am planning on sleeping it off this weekend. That way I will be awake for the IPD review on Monday," I respond.

"Excellent that you will be there," says Dr Cz. "One of these days you will be up and presenting a project of yours for review, so good for you to see Atul perform."

I nod, hoping that I will be able to work up my energy levels in a few days. Right now, I am somewhat discombobulated. Maybe it is the jet lag. Or maybe it is a sort of grieving process – now that my assignment in Europe is over. It really was a fantastic experience. Funny that the things that I was anxious about when I left – like finding a place to live in, or dealing with the languages – turned out to be the easy part. Everyone speaks English, and the Institute had a lady who helps people like me find apartments and with all sorts of other relocation issues. She was excellent.

The hard part – at least for me – turned out to be the weekends. Come Friday 5:00 pm, the good 'citoyens' of Switzerland lock their offices and go home to their very private lives – leaving me, well, alone. Due to my budget limits, the studio I rented there was a bit depressing – somewhat dank, dark, and cold. Hanging there by myself was not nice at all, so after finishing off the loose ends from work, and whatever chores I had, I needed to get away. Rain or shine – I had to go out – even though for half of the year there was more rain than shine. I have discovered that whenever the temperature drops below 50°F (10°C), a drop forms at the end of the nose of this California girl, and she feels thoroughly chilled. Hence the extra clothes that I will never wear again. I, of course, explored all the cities – first in Switzerland and then the rest of Europe. But this was scary – at least to begin with. Wandering around alone… I must have walked a million miles trudging through the admittedly wonderful streets of Lausanne, Geneva, Nyon, and Montreux, at first, and then

Bern, Luzern, and Zurich. And it was only the tiredness that finally forced me to face up to that ultimate fear of mine – going to the restaurants alone. I am fine with it now – and happily enjoy going solo to any café or restaurant any time of day or week. But back then, this was difficult for me, and only the desperate need to sit down and rest, and to use the bathroom, made up for the embarrassment of being alone. In the early days, I found that it felt least awkward in the cafes of railway stations. Now I know that these serve the worst food and attract the worst clientele – but back then, it was somehow less scary. Maybe it is their transit nature. Funny that… But after a while, I got used to it, of course, and now very much enjoy exploring – alone or with Mariano.

Mariano did as he said, and he came to visit me in Europe a bunch of times. It is great that his job allows him the flexibility to work remotely, so he could come and do long weekends with me. It was somewhat snug in my studio – but I found that I actually liked the no privacy bit – but only with him. It was wonderful spending time together – just the two of us – alone in Europe. Late mornings on cold Sundays snuggled in the tiny bed in my dank studio… Followed by those yummy croissants and coffee from the local boulangerie… Picnics in the park by the lake on the sunny days… Evenings walking around and window shopping at the unaffordable restaurants and the high-end watch and jewelry displays, the touristy bijouteries… In all that time with just the two of us, we got to know each other really well, too. And, of course, we have also done many trips together. EuroRail is the best thing ever, and we visited just about all the countries. We ended up using Dario's place in Kaiserslautern as a base for exploring the northern parts. Things *are* expensive there, so this helped. And it turns out that even Dario – my protective big bro – likes Mariano. The three of us – sometimes four, including Inga, Dario's girlfriend – did some places together. That was really great. I think that what they say – about not really knowing a person until you travel with them – is true. You get to quickly discover the various foibles, phobias, and peculiarities that we all have but have learned to manage in our normal daily lives. For example, Mariano is slightly claustrophobic – given his dislike of metros and subways. Anyways, between Mariano's visits to Europe and my biz trips to the USA, we have managed to spend a lot of time and have grown really close. And somewhere in there, without even realizing it, I have gone from liking that man to loving that man. Mariano *is* good.

But it wasn't all fun and games there. I really did work very hard – with prof Doganis to develop his methodology, with various EDA companies to get their tools to mesh with the methodology, and with the San Diego team to get the input data and files. With the 9-hour time zone differences, many days there were 12- to 14-hour marathons. And with Dr Cz doing his thing in the USA, pushing the EDA companies, I did end up shuttling to the various technology centers in Europe – Grenoble, Cambridge, Leuven, and Munich. In a way, it was quite disappointing to visit these great places only to spend the whole day – sometimes two – locked in some meeting room, trying to sell them on the idea of supporting the PathFinding methodology. However, I found that I enjoyed explaining the methodology and talking up the potential value of the approach – converting a roomful of skeptics into believers. I suspect that Dr Cz was right about the advantages of being a woman with a RoCo badge – but only as far as opening the doors, so to speak. The business of converting the audience was all me. Ha! I slayed it. He – Dr Cz – who often patched in to these meetings by phone, said that I am a natural business development/marketing person. I find that idea almost abhorrent – I am an engineer, not some slimy sales person, but… These meetings were exciting and gratifying – and frustrating.

'Gratifying' because I could just see the mind-set of the audiences shifting through the course of the meeting. They start quite skeptical – probably taking the meeting just because it is RoCo, because it is a bad manner to refuse to talk with a lady, and because Dr Cz pushed their VP. In the meetings, their attitude starts out closed and defensive – arms crossed or busily pounding on their keyboards – probably thinking that this pesky girl has nothing worthwhile to tell them. And then, after a while, you could just see their interest perk up, the arms uncross, the laptops close, and they start listening. And pretty soon they are asking questions and even contributing suggestions and ideas. We end up brainstorming and often coming up with some interesting possibilities. Sometimes the discussions would go through, and occasionally well past, the impromptu dinners. I really loved doing that!

And 'frustrating' because, well, the Gartner 'Hype Cycle' Curve seems to apply nicely to the process of engaging the EDA companies.

Source: Wikipedia https://en.wikipedia.org/wiki/Hype_cycle

We have the initial discussions, and we – mostly the technical community – get all excited about the possibilities of the technology and scale the 'Peak of Inflated Expectations'. Then some bean counters or managers start asking questions about target price, market size, and other business questions – and many fall into that 'Trough of Disillusionment'. With some of the companies, we worked past that and climbed the 'Slope of Enlightenment'. But many companies we talked with got stuck in that Trough of Disillusionment and decided not to pursue the methodology. Understandable, I guess. It does not take a lot of analyses to conclude that a general-purpose PathFinding flow – although very valuable and powerful – is not the kind of a capability that would sit on every engineer's desk. So in order to make business sense, the price per installation, for the few that would sell, would have to be huge – which is probably unacceptable to the user community. The companies that we did end up collaborating with tended to fall in one of the two categories. They were either a small start-up who is hoping that RoCo's endorsement will win them a VC backing or maybe even direct investment from RoCo, and as such they were interested in morphing whatever technology they had to fit the PathFinding opportunity. Or they were an established EDA company who saw PathFinding proposition as an incremental avenue to sell some existing technology of theirs.

Either way, we did end up with a few point tools customized for PathFinding objectives. It is not a push-button flow, and expert interference is required to set them up and use them in realistic PathFinding studies. But the tools we got do make these studies more palatable and practical – in terms of the implementation effort required, the turnaround time, and the quality of the results. When I zoom out and think about it – we can now do a bunch of what-if studies and trial designs to get a good insight in the implications of a given technology option for RoCo products. That is so awesome, and I am quite proud of what we have built.

"Jasmine… Jasmine…" says Panchali, patting my back.

"Oh. Oops. I phased out there. Sorry… What?" I ask, returning to the here-and-now in Duffy's.

"Dave was asking about the multi-physics tools," repeats Panchali based on whatever was the conversation I missed.

"Oh. There is a lot going on there. Why don't we get together next week, and I will bring you up to speed? I guess jet lag is catching up with me… Guys, thanks for the beer, but I am going home and sleep," I say, draining my glass and getting up to go.

"OK, Jasmine. See you next week. Sleep well," they say, and I go home. Nice to be going back to my place here – with all that roomy space, my cozy place, and my comfy giant bed.

Chapter 19
Process Transfer Case Study (IPD)

It is Monday afternoon, and we – almost all of the ATI group – are gathered in the Oak Tree conference room for a Phase-3 review of the IPD technology. This is a formal review, and a number of people from the product delivery groups, as well as several VPs, are here.

We have established a structured mechanism for tracking a technology. That is, the 'Phase Review' mechanism – to demonstrate technology maturity to a fixed set of criteria, as it is evolved from the initial Paper Evaluation (Phase 1) to Proof-of-Concept stage (Phase 2), all the way to fully Qualified (Phase 5) stage. The idea is to have a structured discipline for assessing all new technology concepts and progressively down-selecting the candidate technologies for further investment and evaluation. The way Dr Cz puts it is that the Phase Reviews are a succession of 'feed-it-or-kill-it' decisions, to prevent 'starving pet projects' which, he claims, only sap the energy of the organization and produce nothing. This is a Phase-3 Exit review, and the requirement is to demonstrate the intrinsic technology characteristics – nominal performance, intrinsic reliability, engineering assessment of yield and cost, and the like. This is the kind of information that is based on test chip data. Once the Phase-3 Exit requirements are met, the focus moves to demonstration of the statistical attributes of a technology, which typically requires a product, or a product-like test vehicle, and, of course, more significant volume of wafer runs. So, the Phase-3 Exit decision is taken seriously since it gates the transfer of our learning to the product delivery organization, i.e., the design and manufacturing guys, and leads to a significant jump in the cost of effort and materials.

© Springer International Publishing AG, part of Springer Nature 2019
R. Radojcic, *Managing More-than-Moore Integration Technology Development*,
https://doi.org/10.1007/978-3-319-92701-5_19

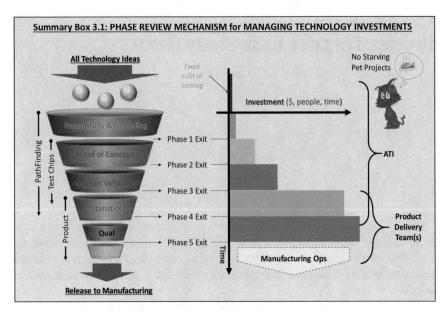

The meeting is run like the test chip reviews – engineers present all the data and analyses, and everybody else picks on it, until there is full agreement that the technology meets all the criteria required to pass the Phase Review. Atul leads the meeting and presents the overview slides that describe the process flow and the overall status. He is followed by a succession of the engineers in his group, presenting various aspects of the technology. He – Atul, aka Darth Vader – is keeping the meeting on track and taking all the notes and action items. I still think that he is a creep – but I must admit that he seems to have done a good job, not only of running the meeting but of leading the whole IPD technology program. It looks really good.

Everything is proceeding according to the agenda, but then, about an hour before the meeting is supposed to finish, Alex – Steve's VP – says: "Can I ask a question that is a bit off the topic – but relevant? I think the IPD technology is in good shape and you guys have done a great job. Thank you and kudos. Atul, please assemble all the data and send it out with the meeting notes, so that everyone here can review it. I am assuming that the technology will pass the Phase-3 Exit. So, I hope it is OK with everyone that I hijack the rest of the meeting," and he looks around the room for any dissention.

Everyone nods – the highlights have already been covered, and the meeting was now diving into details that most of us did not quite appreciate.

"You are the boss," responds Atul, "and yes, I do believe that we have met all Phase 3 requirements."

"Good," continues Alex, "I understand that this IPD technology is now on track to be used in a product. That is excellent. The question I want to ask is what do you think made this technology a successful ATI project? And I am defining success as adoption by a product."

"Excellent question Alex," Steve cuts in, "we have been discussing that very point…".

"Good. Please keep it simple and high level," cautions Alex. "Dumb it down for an exec. I am not after the technical details. I want the top 3 management lessons-learned."

"Yes, exactly," responds Steve, now standing up, "I believe that the principal reasons for the successful adoption of the IPD technology are these..." and he writes on the whiteboard:

1. ~~Lucky~~ *Good Timing vis-à-vis Product Strategy*
2. *Team Versatility and Experience*
3. *New for RoCo – No Legacy to Battle*

"Let me go over each point. Atul feel free to jump in," he says, now facing the room and holding up one finger.

"Firstly, in a way, we lucked out with our timing and positioning. As you know, we have a concept product in mind when we develop and characterize these advanced integration technologies. I must admit in all honesty that the concept product that we were thinking of is not the one that is adopting the technology. The company has recently updated its strategic plans and has decided to move into a new market with a new family of products. It turned out that our IPD technology had some attributes that were attractive for this product family – although we needed to re-engineer the technology to make it fit exactly these new requirements. We were in the right place at the right time with a right baseline solution. Had RoCo not decided to move into this new market, it is possible that our IPD technology would be languishing on the shelf – a solution waiting for a problem. We lucked out that a problem arose just as we were creating a solution. In fact, the characteristic that made the IPD technology attractive to the new product line was almost a by-product attribute of the process – not something we were explicitly pursuing or optimizing."

"Well," cuts in Atul, "we knew that there was a broad technical opportunity with an IPD class of technologies. We were targeting the technology for application-A, and it turned out that application-B leveraged the technology first. It wasn't just blind luck." He seems a bit nervous, I am thinking – took his glasses off to rub his face, revealing slightly bulging eyes staring at Alex. There is a break in his polished non-accented English, with the Indian intonations creeping in his speech. At least he is too busy to be ogling me in his creepy way.

"True," adds Steve, "I do not mean to minimize that at all. I want to emphasize the point that the value is in fact having a rich menu of technology options. The true story of IPD technology is not a case of direct convergence that follows a linear path between the technology idea that we had at the outset, and the actual end application."

"OK, I get that," says Alex. "That is in a way quite typical. If you study the history of various start-up companies, they often end up morphing their initial offering into something entirely different in order to fit wherever they see a market opportunity."

"Right," continued Steve, "but, the key management lesson-learned in this case is that the way we, ATI, intersected the new product line was luck. We do not have a good structured mechanism for keeping all product lines and marketing groups connected to the technology teams, and to whatever ATI is working on. The only reason that the new product team heard of our IPD technology is serendipity. We needed a fourth guy for a game of golf, and lucked out that the manager of the new product line is a friend of a friend... We got talking on the green - and the rest is history. The point that I want to make is that it should not be a matter of luck. We need to fix our management practices and come up with some kind of a structured mechanism that connects ATI with the rest of the company - on ongoing bases. At one point in the past we had an Advanced Technology Steering Committee that fulfilled that role. But it has dissipated by now. Everybody is busy, key people were re-assigned or left the company, new fires came up, and so on. So, now we have nothing beyond an ad-hoc personal network. In fact, I believe that if we had such a mechanism in place, we would not have had to scramble the way we did – to tune the technology to meet the new product requirements. We need to fix that. To me that is Lesson-Learned-#1."

"Aha," says Alex, typing a note on his phone, "I see. Good point Steve. Let's talk about that in my staff meeting and then take it up with the rest of the engineering organization."

"Right. That brings me to the second point," says Steve, holding up two fingers. "We managed to successfully morph - to borrow your word Alex – the IPD technology to meet the new product requirements. And the reason we could do that on the kind of insane schedule dictated by this product opportunity is that we had an experienced and diverse team of engineers who has spent last couple of years massaging this class of technologies. IPD test chips are relatively cheap – so we ran around 10 test chips per year for the last few years, honing not only the technology but also our skills. We got to be really good at tuning the process, characterizing the samples, analyzing the data and implementing the corrective actions – all very quickly. We are good at rapid assessment of the impact of a process change on product characteristics – especially for the specialized types of products that tend to use IPD solutions. We have developed and deployed a set of tools and methodologies for bridging the worlds of process technologies and specialty product design. We have well established and oiled relationships with the right partners in the supply chain. All the bits and pieces were in place... So, when the opportunity with the new product came up, we had a pre-trained team, with the right skill mix, equipped with the right set of tools, willing and ready to morph the technology to optimize product performance and meet the eye-watering cost targets. Since it is unlikely that we will be able to anticipate the exact product needs 2, 3, 5 years ahead of time – not in our business – we need to be ready to do this technology morphing as we converge towards product intersection. Meaning that in order to successfully intersect a product with a new technology opportunity we need to be nimble and able to tweak the manufacturing process and/or the design quickly. And, in order to do this, we need to have the right team in place. So, in my mind, the key lesson learned is that to be successful with intersecting a product with disruptive technologies, we need to maintain a versatile multi-disciplinary team of ... well, we call ourselves 'mutants' … Experienced people who can cover a spectrum of skills between process and design. Not specialists – but integrators. Call it a SWAT team of mutants. To me, that is Lesson-Learned-#2. What do you think Atul?"

"Absolutely agreed," says Atul in his basso voice, "it was not that we anticipated the need to morph the technology in the last minute – it is more a matter of evolving all the required skills and practices while we were doing the development."

"Interesting," says Alex, "so what are you suggesting that the management does about it?"

"Well, simply, that with disruptive kind of technologies – such as the More-than-Moore integration that we have been looking at - it is important to maintain pockets of resources with broad interdisciplinary skill mix, and to task them with taking a technology through the last steps of development towards product intercept. This is different than for the standard - let me call them 'incumbent' or 'mainstream' More-Moore technologies - where there is a lot of historical precedent for making all the usual process and design tradeoffs. So, you cannot develop this kind of a disruptive technology up to a certain point, and then just toss it over the wall to the product guys. They are just not built to morph the manufacturing process – wrong kind of DNA. Similarly, you cannot toss it over the wall to a team of mainstream process engineers. They are not built to comprehend all the implications on this type of technology on design – again, wrong DNA. So, you need a SWAT team of mutants to morph the technology to mate with product needs."

"I see," says Alex, "yes, some companies actually have three kinds of groups for technology development – the development guys on the front end, the manufacturing guys on the other end, and process-transfer guys in between. So, you are suggesting that for disruptive technologies we should consider having a process-transfer team, but with a suitably broad skill mix?"

"Yes. Such as the ATI team here. In my mind the IPD product intersect would not have happened if we did not have such a team in place. That is Lesson-Learned-#2," confirms Steve.

"And," he continues, now holding up three fingers, "the third lesson learned is a sort of derivative of points 1 and 2. The IPD technology is new to RoCo. The adopting product line is also new for RoCo. Consequently, there was no legacy to deal with – on either side. No past technologies or prior designs that we have an investment in, or want to re-use. No prior biases among either the process or the design community. This then gave us a clean slate and an open-minded approach from all people to make an objective decision. Had there been a legacy technology or product to deal with, I suspect that we might have gotten all wrapped around the axle, worrying about the differential risk associated with using a new technology, or design re-use, or what have you. So, in my mind, this lack of legacy actually facilitated the adoption of the IPD technology."

"Are you saying that we should focus ATI on technologies that are targeting only some products-to-be rather than existing product lines?" cuts in Alex. He is obviously listening and paying full attention.

"No," responds Steve, "I am saying that the barrier to adoption is lowest with new product families. That is not to say that the barrier to adoption is insurmountable with existing product lines – it is just higher, in my opinion. And therefore,

I am suggesting that we ensure that ATI is tightly coupled to Marketing and Biz Dev activities."

"Oh, I see," responds Alex, "so should we make ATI a part of the Marketing organization – and bake that tight coupling into the organizational structure?"

"No no no," reacts Steve, rather forcefully, "we *are* an engineering group, and belong in Engineering and Ops Organization – not only by nature of who we are, or so that we can address the existing product lines, but also to leverage company's footprint in the supply chain. This is essential for our strategy of collaborating with the supply chain partners. We need to ride the coat tails of our mainstream products and the associated engineering organization, and must not be perceived as some R&D adjunct to a Marketing Organization. What I am suggesting is that the mechanism for connecting ATI to the rest of the company – the Lesson Learned #1 point – should be especially focused on linking ATI with the early Market Development and Architecture activities. That way we can synch up our technology development activities with their product plans and maximize the probability of product adoption. Our More-than-Moore integration technologies need a runway that is probably similar to the kind of runways that Market Development and Architecture guys think of. This gives us a chance to intersect the product during the early definition and development phase."

"I see," responds Alex, "maybe some kind of a dotted line?"

"Well, maybe. Or a sponsor-VP who would take it upon himself to do the connecting... or some kind of a steering committee that comprises ATI and Market Development and Architecture guys. In my mind, Lesson-Learned-#3 is that we need to ensure that ATI is coupled - on ongoing bases - with the early architecture development activities, so that we can get a level playing field for both disruptive and incumbent technologies. For both, the existing and new product lines."

"Got it," responds Alex, "let's talk about that in my staff meeting too. This is valuable insight and we need to leverage it somehow."

"OK, agreed," responds Steve.

"Excellent. Thank you guys. Good meeting – and excellent work on the IPD technology. Any other thoughts or comments?" says Alex, wrapping up his sideline topic.

Atul looks around, and with no one making any comments, he closes the meeting.

I walk back to my office thinking to myself: "It was a good meeting – even though it was a Darth Vader meeting. I am glad I attended. My time in Europe was coming to an end, and we came back last week just so that I could do this meeting – and I am glad we did it that way … I am not sure whether these Management Lessons-Learned apply to my projects, but they are good points and I should think about it." Meanwhile, I need to go home. I am still a bit jet lagged and so am not feeling like doing any more work – but I do need to finish unpacking and do some home chores. Must keep busy – since I have learned that giving in to the jet-lagged weariness is the worst thing. Coffee and adrenaline are my normal remedy. Yet another hidden gem I have learned during my European stint. This meeting was the adrenaline, and I must have drunk a gallon of coffee. Europeans are right about one thing – coffee there is so very yummy, with that foamy creamy milk. Mmm... "OK, bathroom stop – and then home…".

Chapter 20
Methodology Transfer Case Study (Multi-Physics Stress Modeling)

♪   *Mariano*   ♪
♪   *I just met a guy named Mariano*   ♪
♪   *And suddenly that name*   ♪
♪   *Will never be the same to me*   ♪
♪   *Mariano*   ♪
♪   *I just kissed a guy named Mariano*   ♪
♪   *And suddenly I found*   ♪
♪   *How wonderful a sound can be*   ♪
♪   *Mariano*   ♪

"Geeze, Djaz, singing in the shower? What has become of you? And listen to yourself – calling yourself Djaz?! He has even taken over your self-name. Whatever happened to JLo?"

I *am* talking to myself again – out loud – something I occasionally allow myself when home alone. And yes, I am singing in the shower – a customized version of that cheesy tune from West Side Story. "Funny that mom liked musicals so much when we were kids… Sound of Music, West Side Story, Fiddler on the Roof, My Fair Lady… I think I know all the words to the songs … Must have seen West Side Story a million times when we were little… Maybe it is the cross-cultural love affair bit that attracted her to the West Side Story? But she liked Sound of Music as well – and there is no cross-cultural stuff there... Hmmm…."

And yes, I started referring to myself – in my head – as 'Djaz,' including Mariano's pronunciation. I worried that this is a symptom of allowing him to define me and who I am – down to my name. This has got to be a big no-no in all the women's self-respect manuals. But, hey – it is cute. "I guess he has gotten into my head…"

R. Radojcic, *Managing More-than-Moore Integration Technology Development*,
https://doi.org/10.1007/978-3-319-92701-5_20

♪	*I've got you under my skin*	♪
♪	*I have got you deep in the heart of me*	♪
♪	*So deep in my heart, you're really a part of me*	♪
♪	*And I've got you under my skin*	♪

"C'mon Djaz... You are all over the place... Pull yourself together and off to work you go... But then again – what is a girl to do, when she just got asked to marry the guy she loves... Yes! He did! Last night..."

It is true – and sort of wonderful how it went down. He came back from Peru the day before yesterday – a Sunday. He goes back home every few months or so – to see his family and do some business stuff. He said he would cook dinner in my place – every now and then he likes to do that – and that I shouldn't be late. He made Beef Bourguignon – yummy. We had some wine that he brought back and were just settling in to watch a movie. Netflix and chill. So everything is really nice – but not particularly unusual.

And then he said "Djaz, I would like us to get married."

Just like that! Out of the blue. Made me sit up. He was serious. I mean, I have thought about it – but in an abstract way. One of those things that I thought may happen someday. So, I gave him a hug and a kiss, climbed in his lap, and asked: "Mmm. Nice. But why now?"

"Well," he responded, while nuzzling me, "firstly, because I love you and am in love with you. You Djaz are a wonderful, beautiful woman, and by now, you are also my best friend. I like being with you, and I think we are very good together. And I want to keep it that way. Things are really nice now - as they are. But I decided that I have to catch you before you run off again on some boondoggle or other or get involved with something else new and exciting. I know you – you are a lady on the go. Change likes you and you like change, and when it comes along you get all passionate and are all-in. So, I thought I would squeeze in my bid before anything else crops up. I want to make sure that all plans and possibilities that may come our way are based on '*us*' rather than just you, or just me. And I thought that the best way to do that is getting married ... So, Miss Lopez: will you marry me? Please."

I liked that. And I do love him. So, I said "yes!"

"And," he continued, "I couldn't ask until I got a chance to go to Peru to get this," and he proffered the classic little box "my granny-Farida's engagement ring."

It is beautiful – and looks especially good on *my* finger. Not a gaudy big rock but classically elegant. Feels a bit funny – I guess it will take a while to get used to a ring on that finger and, of course, to everything it symbolizes.

So, yes, I am a bit giddy this morning – singing in the shower, humming silly tunes. Last night we talked about it a lot more and agreed to keep it low key for a while and to celebrate it properly this coming weekend. After all, there are plans to be made, the families to be told, and all that. Afterward, we had a lovely night together. Mmm... I do love that man. But he had to go to work this morning. As do I. But later than usual. Mariano works banker's hours – something I keep teasing him about – and tends to start around 10:00, whereas I am normally at my desk by

8:00. So, this morning we lounged in bed for a while and had a nice long breakfast together. He left, and by the time I showered and got ready, it is already 11:00. I decided to wear heels to work – another way of making the day feel special. Makes me feel more girly. Ha!

And, just my luck – the one time ever that I am late – as I was walking in, I run into Dr Cz, and he asks me to come to his office. So, after dropping off my stuff at my desk, I meet him there, and he says "I need you to cover a meeting for me..." Then he peers at me in a pointed way and asks: "Wait, wait. I am the only idiot around here allowed to sport a grin like that. What's up with you? What are you on?"

I guess my giddiness must be showing, and I must have been smiling ear to ear. So, I told him that I am now engaged to marry Mariano, but that this is still very new and in confidence, and that I don't want him to spread it about for a while yet.

At which he gets up and gives me a huge hug, laughs, and says: "Jasmine! That is beautiful and wonderful. Congratulations. This Mariano of yours - one lucky doggie. Excellent! When is the great day…? And yes, of course, these are your news to spread – not mine." And then after a pause, "tell you what… There is no point in talking to you now – just look at you. Take the rest of the day off and let's meet tomorrow." And he pointedly reaches for his phone and holds it up.

"Who are you calling?" I demand, "you just agreed to keep it private."

"Security… I know you – you will say yes, and then go to your office, open the computer and get sucked into e-mails or something. So, either you leave right now voluntarily and take the rest of the day off, or I am calling security to escort you out," he says with a big smile and a twinkle. Ha! I like Dr Cz.

It is a Tuesday, and everybody is at work. So, I spent the afternoon wandering around Torrey Pines beach and park by myself. Good thing that I keep an old pair of sneakers in my car. It was a nice day, the sea breeze was pleasant, sun was shining, and – of course – my angels were singing. My private personal celebration. Yaaas! Life is wonderful!

The next day, I go in early and straight to Dr Cz's office – but he is not there. So, I go to the coffee room to get my morning coffee fix and run into Panchali there. I am trying to contain myself – but do put my hand on the table casually displaying my new ring. She is crafty – this Panchali – and it did not take her long to spot it. "Hey, is that what I think it is?" She asks, eyes wide open, big smile on her face.

"Errr," I am not sure whether to tell her. We haven't told our families yet, and I certainly do not want to make a big deal out of it – especially at work. "Oh, what the hell… Yes. Mariano has asked me to marry him, but…"

She virtually screams and gives me a big hug, and it is too late to calm her down and keep the cat in the bag, so to speak. There is a lot of traffic through the coffee room in the mornings, and pretty soon everybody is congratulating me, and there is a bit of a gathering there. So much for keeping it quiet.

Then this guy I have not met walks in and asks "what is the fuss here?" while filling his mug.

"John?" yells Dave, "good god, John Williams? Good to see you. Where the 'll have you been all these years ... How long has it been?" and they shake hands and pat each other on the back, bro-style.

"A few since we last met. 20-ish since Burroughs... You?" responds John – clearly, not a man of many words. He is a bit like Dave. Anglo in his 50s. Jeans, worn-out plaid shirt, and gaudy western belt buckle, carrying a big coffee mug that looks like it has not been washed in those 20 years they are talking about.

"I am with the Advanced Technology Integration group – still doing modeling and stuff. And the fuss is that Jasmine here is getting married," says Dave, and then turning to me "Jasmine – this is John Williams. We used to work together in Burroughs, probably before you were born."

We shake hands and I ignore the well-worn comment about my age. John says "well, congratulations ... and glad to meet you." And then to Dave, "I am with the Failure Analyses group here – joined around 7 months ago," and to both of us, somewhat grumpily, "good to see that someone still has the time for life... getting married... and celebrating."

Turns out that there is a major issue with Product Blu, and, apparently, he has been working 16-hour shifts 7 days a week. He says "yields are sporadic and go up and down in seemingly random ways, so we are scrambling to find the root cause. It is costing the company in a big way, so they have assembled a fire-fighting team to debug it. All-hands-on deck kind of scramble with 2 meetings per day – every day... So far, it is looking like a real head-scratcher. All the failed parts that have been submitted to me for Failure Analyses have nothing wrong with them. There is nothing like a smoking hole there. And all the teardowns and inspections that I have done reveal no defects that I can tie back to the observed fails. Nothing. Nada. Every failed part I looked at looks pristine."

"Hmm," says Dave, "there has got to be something. Usually when the yield is sporadic and you guys from FA report No-Defect-Found, the issue is some deviation in the fab process."

"Yes, of course ... I am as old as you and have lived through as many of these fires as you have," responds John somewhat impatiently, "we looked. There is no correlation with anything in the fab. The only signal we are getting is that the version that keeps having the yield issue is the one that uses this new package. But no one can figure out what is it about that version that would cause the failures"

"New package?" asks Dave.

"Yeah... Well – new to Blu. The package has been used in other products. Nothing special, really. Maybe a bit thinner than the standard package," answers John, getting ready to leave – with his giant mug of coffee.

"Ah!" exclaims Dave, "thinner you say? Maybe it is something to do with mechanical stress? If so, you may be in the right place, my friend. Jasmine here has been playing with some simulators that model the piezoelectric effects. You guys should talk."

"Stress? Maybe... Except that there are no cracks, no sign of delamination, and the package meets all the usual flatness standards – no observable warpage or anything..." responds John, clearly skeptical about the possibility of mechanical stress being the root cause of his problem.

I am glad that the topic has shifted to something other than my engagement. But his dismissive skepticism is irritating, so I take it as a kind of a dare and want to prove him wrong. Everybody is used to the symptoms of any issues that involve mechanical stress to be warpage, cracks, and/or delamination. But when we – ATI – looked at stress while we were exploring the Through-Si-Via technology, we found that mechanical stress can also cause a significant change in transistor characteristics (*TBB 3.1). This is the piezoelectric effect described in some of the literature out there. It is not rocket science but is a niche that most people are not familiar with. So, I ask "What is the die thickness? Our modeling show that this is the dominant factor with stress related parametric changes?"

Technical Background Box 3.1: Mechanical Stress Effects and CPI

- *Sources of Mechanical Stress*: mechanical stress can be caused by many phenomena, but the most obvious source is the mismatch of the thermal characteristics of different materials. Materials naturally expand/contract as a function of temperature at a specific rate, as described by the Coefficient of Thermal Expansion (CTE). The CTE for Silicon is about 3 ppm/°C, for Cu about 9 ppm/°C, and for organic substrate materials used in packages and PCBs about ~ 16 ppm/ °C. Thus, at room temperature there is some residual stress in an IC due to CTE mismatch between Cu, Si, and SiO_2 since the typical deposition temperature (corresponding to a zero-stress situation) is in the order of ~400 °C. Similarly, at room temperature there is residual stress between the Silicon chip and the package, since the die attach temperature is in the order of ~300 °C and so on. This stress needs to be relieved and can naturally result in strain – both locally and globally – leading to bending and warpage, depending on the size and thickness of the structures and stiffness of the materials involved and so on. In some instances, this stress can lead to catastrophic failures due to cracks, fractures, or delamination taking place at some interface. Managing stress in a packaged IC caused by CTE mismatch between Si and the package – often referred to as Chip-Package-Interaction (CPI) – is a fundamental challenge and is typically controlled through warpage and planarity specifications, which in turn dictate the choices of process temperatures, sizes, materials, package design, etc.
- *Effect of Thickness*: Silicon wafers, as used in the fab, are about ~1mm thick, and traditionally chips were assembled in packages at full, or near full, thickness. At that thickness Silicon chips are very stiff, and the stress induced by CTE mismatch is relieved through deformation of softer materials – such as solder or the package substrates. However, with many consumer devices that favor a very thin form factor – such as phones – Silicon is thinned down to 100 μm or less, resulting in IC die that are quite flexible and able to relieve stress through local bending and overall warpage (the picture shows a full wafer thinned down to 50 μm – it's a pancake!).

(continued)

Technical Background Box 3.1 (continued)

- *Piezoelectric Effect*: bending and warping of Silicon lead to a change in the electrical characteristics of devices on the chip. Basically, bending the Silicon crystal alters the lattice constant (distance between the atoms) which leads to a change in the electron mobility (basically ease with which electrons move through the material). Thus, packaging very thin die can result in change of IC performance – this is referred to as 'e-CPI' effect. The degree of change in the device performance is dependent not only on the magnitude and direction of stress (i.e., the basic CPI effect) but also on many other factors such as Silicon crystal orientation, device polarity and orientation, circuit design, etc.
- *Managing e-CPI*: the phenomena outlined can all be modeled and simulated but do require a multi-physics approach, i.e., the CPI effects have to be modeled in the mechanical domain and the consequences are then simulated in the electrical domain. See, for example, 'Analysis of the Effect of TSV-Induced Stress on Devices Performance by Direct Strain and Electrical Measurements and FEA Simulations,' Valeriy Sukharev et. al., IEEE Transactions on Device and Materials Reliability, Year: 2017, Volume: 17, Issue: 4.

Note: material in the gray boxes is intended for those who are interested in more semiconductor technology and/or industry background information and may be skipped by those who are not.

"Well, I think that the die is a bit thinner than normal. The thickness of the package and the chip needs to be balanced, so when they thinned down the package, they also had to thin down the die – in order to meet the warpage specs," John responds, hesitating a bit – obviously thinking about it. "Look, we need to solve the problem and at this point will look at anything that is remotely plausible. Why don't you guys join us this afternoon in the War Room and let's talk about it. 5:00 in Elm Tree conference room," he says and ambles out.

So, Dave and I agree to join the meeting to see if our methodology can help. Through-Silicon-Via process works only with very thin Silicon wafers. In fact, Silicon wafers are so thin that they are transparent and sag like a pancake under their own weight. That, and the fact that Through-Silicon-Vias are filled with copper, which results in mechanical stress gradient in a microscopic region around the via, has led ATI to explore the ways of simulating these effects. I collaborated with some of the EDA vendors on this and have been working with Dave for a couple of years now – to develop a multi-physics modeling methodology that comprehends both the mechanical and the electrical effects. So, we now have the software tools and have done some simulations and experiments to validate the methodology.

War Room is a good name for what we found. Elm Tree conference room has been reserved for the team working on the problem. Graphs, pictures, flip charts with scribbled notes and sketches hanging on the walls, and whiteboard filled with lists of action items and random doodles. People – only some of whom I know – looking harried and busy going in and out of the room, with multiple conversations going on at the same time. At about 5:15 this guy walks in and takes the seat at the head of the table, and the room gradually quiets down. He is evidently the lead on the project.

Before he could say anything, John clears his throat and says "Mike. This is Dave and Jasmine. They are with the Advanced Technology Integration team. I asked them to join us today because they have a mechanical stress modeling methodology which may be helpful. I think we should hear what they say and have a look…"

"OK. We have already looked at mechanical stress. There is no sign of an issue. Everything is in spec. So – what is new with your methodology?" responds Mike, looking from Dave to me and back. He gives an impression of being skeptical but is evidently willing to listen – maybe because he is desperate.

"We model the effects of mechanical stress in the electrical domain," I say, jumping in before Dave could say anything. I am feeling like this Mike is going to give us maybe all of 5 minutes, and Dave does have a tendency to meander and usually takes a while to get to the point.

"So?" responds Mike.

"So, we have found that the effects of mechanical stress on transistor performance can be more pronounced than on the usual mechanical characteristics - if Silicon is thin enough."

"So?" repeats Mike.

"So, it is possible that device performance changes of more than 5% to 10% occur before effects like warpage or cracking are observed - depending on all sorts of variables," I say, trying to be as focused and short as possible. This is a complicated phenomenon, and it is easy to get sucked into a rathole debating and describing any of the many dependencies.

"You say this is thickness dependent - how thin?" asks Mike. He is still skeptical. 'Shields up' but at least he is listening.

"That depends, but I would say that somewhere around 100 μm the electrical effects start to be significant," I respond, making a wild estimate based on what we have seen in the simulations in the past. "What do you think Dave?"

"Well, that depends on the magnitude and polarity of the stress, thickness of the wafer, crystal orientation, type of a transistor, construction of the package, circuit sensitivity…" responds Dave, getting warmed up for a good discussion.

"OK," Mike cuts him off, "10% shift in device performance could definitely cause failures. For this package, Blu die is thinned down to 80 µm. So – what do you need to model it?"

Good, I am thinking – the door is open, ever so slightly. "If you provide us with the construction information – things like sizes and thicknesses of the die and the package layers, some basic material properties, and the layouts, we could do a quick assessment and see if the piezoelectric effect is in the ball park… If it is, then we can get more detailed and do a round of better modeling," I respond, thinking that we do not want to have that door slammed shut now, by getting all bogged down in a long list of detailed requirements.

"Good. John, can you provide that information to … err … sorry … what was your name? When can you get back to us with the results," responds Mike, clearly ready to move on.

"I am Jasmine and this is Dave… Beginning of next week. But these will be preliminary. Just a sanity check," I respond. I am winging it now – not having seen the inputs – but feel that this Mike is a black-and-white sort of a guy. For now, I suspect, he just needs a date to fill in some action tracking spreadsheet. To him, this is a sideline that must be explored – no stone unturned – but he is not counting on it for a solution.

So we go to John's office, and he gives us a list of the relevant specs. We – that is, Dave and I – then go to talk to Dr Cz, to make sure that it is OK for us to pursue this; after all, this is not an ATI project. He asks what will not get done if we do this and how long will it take, and after some discussion gives us enthusiastic support – saying that it is the right thing to do for RoCo and that it is also an elegant way of demonstrating the value of ATI. So, I spent the next few days running around gathering all the necessary information about Blu Silicon and Package design and so forth. This is always a challenge because package-related information is in one database, maintained by one set of people, using one set of formats, standards, and acronyms, while Silicon design data is in another database, maintained by entirely different set of people and based on totally different formats and standards. And after I finally got access to the data, I had to write some scripts to put it in the form that can be used as an input for my simulators. We also had no calibrated models for Blu technology, so Dave and I brainstormed and agreed on what would be good values to use. It was a bit better than guessing, since we used a combination of models from the literature and the ones we derived when we did the TSV analyses. Then I kicked off several simulations in parallel to run over the weekend. These simulations take a while – sometimes more than 24 hours. And I wanted to do several runs, using a range of input values, so that we could get a feel for the sensitivity of the outputs to the inputs.

Mariano was a bit miffed that I took the time on Saturday morning to check the status of the runs – since this was supposed to be our engagement celebration weekend. But I reminded him of all the times when he took time to watch some soccer match or other. He mumbled something about having to share me with work – but was kidding. We ended up having a great weekend. In the afternoon we took the ferry to Coronado, walked around the village, had a fine dinner at Hotel del Coronado, and talked a lot about us, the plans for the wedding, the logistics of moving in together, families and friends, timing of various events, and the like. We will go and tell my parents the next weekend and will go to Peru over Christmas to meet his folk. Nice.

The following week Dave and I reviewed the results of the simulations, concluded that they made sense, and prepared the presentation slides – trying to stay focused on the results rather than the sheer elegance of the methodology that we developed. The results are almost qualitative – since the models we applied were based on best guesses – but do demonstrate the power of the methodology very nicely and do have interesting implications for Blu. We felt like proud parents about to show off what our baby can do.

So, the next day, we go to the War Room and ask Mike for about 15 minutes to show our results. He nods, and I go through the background slides quickly and put up our star slide – the results we wanted everyone to focus on: a map of high-risk areas, i.e., areas with largest shift in transistor performance caused by the chip-package mechanical stress, superimposed on Blu die floorplan:

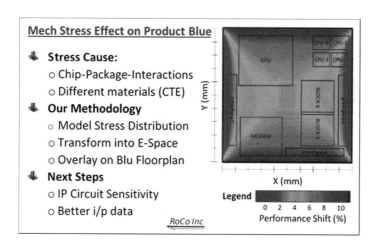

Silence in the room. Everyone staring at the slide, trying to understand how we got there.

Then Mike clears his throat and says: "OK. I think I understand what you guys did. Nice, thank you. So, correct me if I am wrong, but the bottom line is that the Package induces up to 10% gradient in performance along the die edges? That is on top of the usual statistical process variability?"

"Yes," I respond, "with the default values for the models and material characteristics that we used. The sensitivity analyses show some dependence on the input parameters, of course. So, if we re-do the simulations with better input estimates, it may turn out to be 5% rather than 10% spread – but the distribution will be as shown."

"The key issue is not so much the exact value of the input parameters," adds Dave, "but the sensitivity of the circuits to the performance shift and/or performance gradient in the affected region. Some circuits may not care, others may be sufficiently sensitive to fail functional test, and everything else in-between."

Then John speaks up, a bit hesitantly, as he is working it out in his head, "Speaking of test … these results are based on just the chip and the package? Right? It does not include any externally applied stress – like for example the fixtures and clamps used when the part is tested? Right?… These packages are quite flimsy and may be bent during standard test."

"Right," responds Dave.

"Or the potential internal stress gradient that could be induced by a hot spot on the chip? These chips get quite hot during test," adds another guy.

"Yes," adds John, "there could be all sorts of additional sources of stress during active operation of the part under test. This may add noise to the yield data – since the external stress applied to the package during test is not controlled."

I love it when this happens! The roomful of skeptics is now busily brainstorming on all sorts of other potential sources of stress that could be causing the Blu issue. Ha!

"OK," cuts in Mike, after about 45 minutes of the discussion, "My head hurts. Too much info. 5% to 10% shift in performance could very well precipitate the failures that we have observed, and all these other factors may be responsible for the sporadic yield signature we are seeing… Jasmine, Dave – what do you need to make a conclusive analysis?"

And that is what it took. Following that, Mike directed this or that, and we were buried in a tsunami of inputs and got all the help that we needed from anyone that he recruited. I guess a yield bust that costs the company money does trump anything else. For the next month or so, I had the simulator running 24/7, and it turns out that stress-induced performance shift *was* the issue with Blu. With the full understanding of the root cause, the team identified a few 'Band-Aid' fixes to manage the problem until a new package was designed and qualified. Dave and I received a 'Diamond RoCk' – a company award – in recognition of our contribution. Ha! I guess it is true what they say – that the glory goes to the firefighters solving some known product problem, rather than the engineers avoiding the problems in the first place. Dr Cz and Steve were also delighted with ATI getting the visibility from other teams. I will have to remind them of this during the next performance review. That bonus will help with the costs of the trip to Peru, the wedding, and stuff.

A month or so goes by. Mariano and I went up to LA to meet with mom and dad. They were wonderful – delighted. And of course, they loved Mariano. Everybody likes Mariano. He does know how to work the room, so to speak. The butt-kisser that he is, he brought flowers for mom and a bottle of Pisco for dad. In a few hours, he had them eating out of his hand, so to speak. Mom was so excited that you would think that *she* was getting married. And dad opened the bottle, Mariano made some

Pisco Sour drinks, and pretty soon they were arguing about soccer like old amigos. It was all really sweet.

And then, the following week, I get a call from Mike asking me to help out with Blu-Lite, a derivative product that he was driving. I told him that I would of course love to, but need to clear it with my manager. But when I brought this up with Dr Cz, he said "That is excellent. Kudos to you and Dave. But, on the other hand, we cannot have random people borrowing ATI resources for product delivery tasks. That is not what we are supposed to be doing. Why don't you call a meeting with Mike and me, and let's talk about it…"

It took a while to find a time slot on when everybody is available – this Mike seems to be a busy dude, and Dr Cz is not that much better. Turns out that Mike is a product manager – apparently one of the very influential ones – responsible for Blu product family and that they are now designing a derivative chip targeting lower-cost implementation. And Mike said that having seen the value of our methodology in dealing with Blu yields crises, he wanted to make sure that we avoid similar problems with Blu-Lite. Ha! It is wonderful that he is now a convert and wants to be proactive. So, he wanted us to do the analyses ahead of time to make sure that Blu-Lite design is immune to stress issues. Dave and I explained that this could take us a couple of months, and Dr Cz emphasized that direct support of product delivery is normally not a part of ATI mission. So, after some discussion, it was agreed that we should identify a new 'owner' of the methodology in product delivery organization and then assume a consulting and supporting role.

So, Dave and I started doing the rounds, setting up meetings with all sorts of people from all sorts of groups. We met with Silicon designers – who said that (a) they know nothing about package design and (b) they need an EDA tool enabled for a sign-off, i.e., a tool that would contain the pass/fail criteria for acceptable stress-induced performance shift. Then we met with Package Designers – who said that (a) they know nothing about Silicon design and (b) they need a pass/fail spec for mechanical stress that Blu-Lite performance can tolerate. We met with product and test engineers – who said that (a) they are busy supporting existing products and (b) they know nothing about Silicon or Package design or stress. We met with Package Process Engineering – who said that (a) they need a stress spec limit ahead of time, so that they can engineer the material characteristics for the package, and (b) they do not have access to in-process temperatures, which is one of the key factors for working out the residual stress. We met with Silicon Process Engineers – and they said that (a) models for sensitivity of transistors to stress are not provided by the foundries and (b) it cannot be measured directly without specialized full flow test chips. And so on. There was no one who felt that they either had the responsibility or the expertise to assume the ownership of our methodology. No one. So, we called another meeting with Mike and Dr Cz to explain the conundrum.

Dr Cz's reaction was almost philosophical, and he said: "Well, I guess if we were smarter we should have anticipated this. ATI is a group of cross-disciplinary mutants and normal delivery organizations are silos full of specialists. I guess it makes sense that no one in existing organizations can assume the wholistic ownership…"

And Mike on the other hand was a bulldog and said "yeah yeah yeah. I don't care. I need this work done. It is important for revenue generating Blu-Lite," with emphases on the 'revenue generating' part. "Cz, I think you guys have to do it for Blu-Lite. I need Jasmine and Dave to do their thing. You can work with whoever you can find to assume the ownership, but until then, you guys are holding the bag."

I was shook by all this. It looked like our methodology – proven to be good – is doomed to be an orphan. Product delivery guys cannot adopt it because they don't have the right skill mix or broad enough charter, and ATI cannot use it to help in product design. WTF? This makes no sense! I am very proud of the methodology – especially now that it has been proven – so this is frustrating. Surely, the right thing to do is to ensure that the company products are good – and I said so, even though I know it wasn't my place to argue this management decision. At which Dr Cz, sill in his philosophical mood, responded "Well, the reason ATI exists is so that the company has a pocket of resources that are immune to product pressures. Product pressures are always more important and more urgent, and if we allow them to divert these resources, we will never do any advanced technology development...We would not have the methodology to begin with if…" And then, after a pause, he adds "my mentor once told me that management is all about dealing with ambiguities and living in the world of multiple #1 priorities. So, we will just have to manage this too. In the long run this is a strategic issue, so I think we will kick it upstairs to Steve and his Steering Committee. Meanwhile, Mike, yes, we will do what we can for Blu-Lite. Steve will probably need your help with the Steering Committee though – to find the permanent owners. And maybe we should visit the EDA guys to see if they can be induced to make a push-button tool."

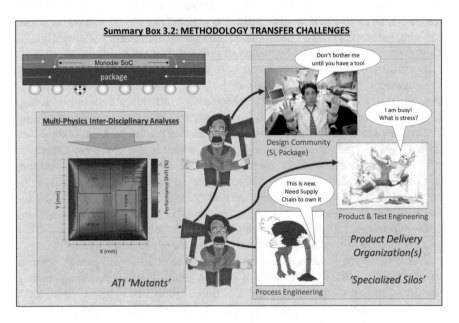

And that is the way it was. Dave and I did all the work, of course, and helped Silicon designers to make a more robust floorplan for Blu-Lite die and Package Designers to do a layout that would spread the stress more uniformly. Funny how the work that we did a couple of years ago, on Through-Si-Via technology, is now yielding results used in actual products that have nothing to do with TSVs. The thin wafers used in the TSV technology were an extreme case that ended up being a sort of a 'canary in the mine' that alerted us to performance sensitivity to stress effects – and we now have this cool modeling methodology. Interesting. We are also working with an EDA company to develop a tool that models the phenomena. However, by definition, the methodology involves multi-physics and requires multiple disciplines in multiple domains to calibrate the models – which will still probably be too complicated. So, we are no closer to finding permanent owners of the methodology, and Dr Cz keeps grumbling – albeit in his good-natured way – that the stress modeling has become an anchor around our neck, since it keeps sucking ATI resources away from our core mission. It is a conundrum. On one hand helping the product guys is the right thing to do, and I do want to keep the methodology alive. On the other hand, it is not our job.

Chapter 21
Concept Transfer Case Study (2.5D Integration PathFinding)

We are back from Peru. We went there to meet Mariano's family. It was a bit of a hurried experience – since we could stay just 1 week – because both Mariano and I had used up our vacation time for all those side trips in Europe. And, it turns out Mariano was almost broke and that, being mostly on commission, he needed to work. He said he used up his savings on all the trips to Europe but that he had no regrets whatsoever and felt that he made an excellent investment – since he now has me. Ha! So, I paid for the trip and lent him some money to help with his rent and stuff. I could see that he felt bad about that. We still need to talk through how we plan to handle our finances – especially after we move in together. But I thought that helping out is only fair – since he spent all his money on me and our relationship. The visit with his family was great. A bit of a whirlwind – so much family, so little time. And they all talked Spanish so fast – and so much – that I missed a lot of what was going on. The language we spoke at home, when I was little, was English, and the bit of LA-Spanglish that I picked up was not adequate to keep up with all the conversations – most of which, it seems to me, ran over each other. But they were all really nice. Lots of hugging and kissing and touching. Too bad that his dad was not well – the poor man has Alzheimer's. Very sad. But mom, sister, uncles and crazy aunts, cousins, and what have you were all awesome. And it seems like everyone there knows Mariano. We walk down a street, and inevitably we would bump into someone he went to school with, or played with, or was someone's cousin, or a friend of a cousin's neighbor, or something. Altogether, it was excellent – although I feel like I gained 50 lb. with all the delicious food there. We will definitely have to go again – many times. I loved it. Anyways – now we are back – it feels like a month, but it was actually only a week.

So, it is Monday morning and I go in early – to grab Dr Cz before all his meetings – to catch up on whatever I may have missed. He of course, asked about Peru, and the trip, and the dreaded mother-in-law, and so on. Then he gets up, closes the door – indicating that he wants to talk about something serious – and says: "I am glad you are back. I have things I need to talk to you about... It's like this. We met with Bill Mao. He is a VP of the Architecture team. He was saying that he has a

© Springer International Publishing AG, part of Springer Nature 2019
R. Radojcic, *Managing More-than-Moore Integration Technology Development*,
https://doi.org/10.1007/978-3-319-92701-5_21

strategic concern that with each technology generation our chips seem to be getting bigger in size. Apparently, market requirements for the high-end products keep growing, so that even though we scale everything with each new process, the content of the chip keeps growing, and the end die size increases rather than staying constant or decreasing. More CPU's, bigger GPU, more radios, more specialized accelerators, and what have you."

"So? That is all good, isn't it?" I ask.

"The problem is," he continues, "that the price that the market can bear for these ICs supposedly stays more or less constant. That goes back to the end consumers – people like you and me – who are not willing to pay much more every time they go and buy a new device. So, the trend of growing die size is alarming because the price has to stay constant."

"So? Hasn't this always been the case?" I ask again.

"Well, the growth in die size is not because we are integrating other chips functionality on our die – which would reduce cost somewhere else. The problem is that it is growing because we keep adding new functionality, new performance and new features – which naturally should be reflected an increased price. In addition, the wafer manufacturing cost is expected to increase for the next few generations of process technology much more rapidly than was the case in the past. This is due to the double or quadruple patterning – i.e., the lack of a practical Next-Generation-Lithography solution – and the general increase in the complexity of the process technology. Yet another symptom of the end of Moore's Law. However you turn it, the die size and Silicon cost-per-unit-area keep going up, but, apparently, the market will not bear the increased price," he explains.

"Well, that does not sound good. Sounds like a squeeze on our profit margins?" I ask.

"Yes. So, he – Bill Mao – has some long-term strategic concerns. Steve suggested that a possible way of reducing the cost structure is to split a big die into two smaller die, and then to re-integrate them at the package level. As you know – that is one of the value propositions of our favorite 2.5D and 3D Integration technologies," Dr Cz continued.

"Absolutely" I say, now more interested. If this ends up being that 'window of opportunity' that we were looking for – an opening to drive our technologies into a product – it would be so very cool.

"So, long story short, we agreed to do a joint engineering study – resourced from Bill's Architecture team and ATI – to look at this possibility. This would be the most comprehensive PathFinding study we ever did – and it would be for a real product, rather than a test case," he concludes.

"That is awesome," I respond, definitely getting excited. "When do we start?"

"Soon. We first need to define the scope, schedule and all the other parameters of the study. Then we need to define a project and staff the team. And then we need to execute – all the simulations, trial designs and analysis we have been talking about for the last couple of years... And – this is what I wanted to talk to you about – I would like you to lead the project," he says.

"What? Me?" I ask – surprised. I was not expecting this. Whoa!

"Yes, you. You are closest to the PathFinding methodologies and have the most experience with the knobs and tradeoffs required to do the simulations. Plus, you know the capabilities and characteristics of the process and packaging technologies. So, in short, you are a mutant who has been playing with these kinds of studies for a while. And, I am sure that you can get all the help you will need from the Architecture guys for the software and hardware design aspects that you do not know… And, I believe that you can do it…," he says pointing at me in his Uncle-Sam-Wants-You poster kind of a way. Then he peers at me and adds "The question for you is (a) in view of your upcoming nuptials will you have the time, energy and focus for something like this, and (b) do you want to lead a project." He says "nuptials" in a funny way; I guess trying to do a snooty British accent.

"Errr. Duh. Yes, of course, I want to lead a project. We have not set the exact date yet – but it will be sometime in the summer. And this sounds like such a cool project," I respond.

"Wait. Slow down, grasshopper," he cautions, "think about it. It will be a lot of work, and it will put you in a role that you are not familiar with. Leading a project of this size – and profile – would be something entirely new for you. I think you can do it – but you need to want it and be sure of it in your head and here," he says tapping his chest. "Doing it would give you visibility – but depending how the project goes, it may be good or bad visibility. Think about it."

"OK, Dr Cz, thanks. Umm…thinking… thinking… I want it!" I say, "Consequences be damned – let's do this!" I am excited about the project and the prospect of being the lead. And I have learned that I know what I know, and I know that what I do not know, I can learn. Someone said that a while back – and it stuck with me. And I actually believe it. I guess I am a big girl now, and I want to be the lead on this project.

"Tell you what. Why don't you first go and talk with Sam S. He is the point guy on the Architecture team and would be your counterpart. Why don't you guys get together and brainstorm a bit, and then give me your commitment. This is important for us – for ATI and for RoCo – so we have to be sure," he says, "let me know how it goes."

So, for a week or so, I conducted my version of shuttle-diplomacy. Chatting with Sam S plus a number of other key people and getting a feel for the scope of the effort, the constraints, the people who will participate, and whatever other concerns there are.

When I talked with Mariano about the prospect, he said "See! I knew it. My window of opportunity was in reality tighter than I expected. I am glad to have slipped that ring on your finger while I still could," and added jokingly, "you cannot back out now, Mrs. Llosa." Not that I would want to. That would be all wrong – having to choose, I see no reason why I cannot have both – the project and the man. And one of the best things about Mariano is he also saw no reason why we could not do both. And I really liked it when he said 'we'. This marriage gig may not be bad after all.

By the time that the formal kickoff meeting rolled around, I was ready – I had a pretty good idea of what needs to be done and have met all the principal contributors whose help I will need. The kickoff meeting included Bill Mao and Alex – they are the executive sponsors – plus Steve, Dr Cz, Sam, and a number of people from the architecture team, as well as the 'usual suspects' from ATI.

Following the normal preliminaries, including the formal introductions and the definition of the project intent, Bill Mao – an older Chinese gentleman, somewhat taller than I expected – says, in a terse, methodical way: "Look, the company needs us to tape out our next product on schedule. So we have a plan-of-record – a single die SoC implementation. At any one time, there can only be one plan-of-record. This Split-Die architecture using 2.5D SiP technology is an unproven opportunity, and therefore cannot possibly be the official plan-of-record. So, it is a Plan-B. We cannot put Plan-A on hold, so the Architecture team will drive to it. I believe that there is a strategic problem and an opportunity with 2.5D technologies, so we will provide whatever support is required. But it makes sense that ATI drives Plan-B, while the Architecture team focuses on Plan-A. Given the overall product development schedule, the point of no return when we have to pick either Plan-A single die SoC or Plan-B split die SiP, is a quarter from now – so we have a tight window of opportunity. You guys have three months to do the analyses. Then let us get together, review the results, and make a data driven decision."

"That makes sense," responds Steve, "we appreciate the support. I am asking Jasmine here, to be the lead – with Cz's help. Jasmine, please set up a weekly meeting with me and the executive sponsors to keep us informed, in parallel with the project execution meetings and all the other activities you will have. For the first one of these show us three things," and then he counts off: "Firstly, the project scope – i.e., the specific results you will deliver 3 months from now. Secondly, a delivery plan – defining the schedule for the key tasks, and their dependencies. And thirdly, a resource plan – identifying the people you need and highlighting any conflicts that you anticipate. Use us, the management, to bless these plans and to resolve any conflicts. There is not a lot of time, so let's go go go!"

Wow! A tall order. And none of the touchy-feely stuff. I have received my marching orders. My butt is on the line to deliver – no matter what. A lot of work, a lot of meetings, a lot of stuff to learn, a lot of analyses and reviews, a lot of things to do and people to see. But I volunteered, so I can do this!

Then, about a week later, just before the first of these exec review meetings, Dr Cz comes by my office and says "Jasmine let's take a walk outside."

What? This is unusual. Normally he just says what he wants. Why is he taking me outside – is this like the proverbial being taken behind the shed for a whipping? A knot forms in my stomach. What have I done?

We go out, and he says: "Look, what I am about to tell you is not the official line. That is why we are outside – badges off. I am now not an agent of the company – just a colleague or a friend."

Uh-oh. This is definitely not helping my growing sense of anxiety. "OK Dr Cz?" I say haltingly – waiting.

"Sam is off the PathFinding project. He went to see Bill and told him that he is not prepared to work on this project with you as the lead," he says.

"Uh-oh... So, are you asking me to step aside and give up the project leadership? Already? C'mon... Because if you are, I am...," I hesitate, maybe somewhat warily and defensively, feeling like my prize is about to be snatched away. A bubble forms in my throat.

"No no. Hold on and listen." He cuts in, "Bill got together with Steve and me, told us about Sam's reluctance to work for you, and said that there are four combinations for this project: you leading a team including or excluding Sam, or Sam leading a team including or excluding you. He – Bill – said that given Sam's reaction, combination-1 is now off the table, and that whereas he felt that Sam is the right guy, no one is irreplaceable and team harmony is paramount. But, since it is our project he wanted us – ATI – to make the decision. At which Steve indicated that ATI is a small team, and that we don't really have a backup for your skill mix. He suggested that I could take more of an active role, if necessary. But then he said that taking the leadership from you at this point would have not only legal, but also ethical implications, and that we cannot let some individual's personal preferences dictate our management decisions. And then – you would be proud of Steve – he said that taking you off the project at this stage, for no cause, would be just plain-wrong. We talked about it and agreed that we should carry on with you as the lead, and that Bill will assign Max to the project to replace Sam."

"Oh, phew... Great. Thanks. I appreciate the confidence. If you guys feel that Sam would do a better job, I would of course step back. I would hate to, but I will do whatever is the best for the project. I am glad that I don't have to, though... So, why the secrecy – this walk outside...?" I am relieved – but still puzzled.

"Well, the official line is that Sam has just been assigned to a different project. No big deal. This happens all the time. End of story. But I thought that you should have the full background. I think that we would be setting you up to fail if we asked you to play without a full deck, so to speak. Sam is influential in the Architecture team and you need to know the whole story and be sensitive to wherever he is coming from."

"OK," I respond, "Thanks. Now what?" I am relieved, but there is something else here, so still a bit wary.

"Nothing... We press on. But, it may be worth thinking through where Sam is coming from and understanding *his* perspectives. So that you are prepared for whatever his reactions may be to your project and the results," he says.

"Yes, that was my next question. What is Sam's problem with me?"

"Well, I don't really know," Dr Cz says, "he is more senior than you are – but I am pretty sure that in the past he did not have a problem working on teams led by engineers a lot less experienced than he is."

"So, is it because I am a woman?" I ask, with the ugly reality slowly dawning on me. I did not get any vibes like that from Sam. I mean he was a bit terse and abrupt, but I assumed that this is just the way he is. Not the warm and fuzzy kind of a guy. Stern, short, and prone to use of monosyllabic responses. There are people like that – without necessarily being sexist. On the other hand, maybe he was that way just with me and with others he is this gregarious, friendly dude? Maybe this was his way of protesting that I am a woman?

"Possibly. I do not know. But – assuming that this is the reason – we need to manage around it. Right, wrong or indifferent, we cannot pretend that Sam does not matter, or that he will disappear somehow, or that his attitude will magically change overnight," he says.

"Well, I cannot change the fact that I am a woman. Nor would I want to. So, what the hell am I supposed to do? If his contribution is important, then am I handicapped from get-go, with him off the project? Am I now set up to fail because I am a woman – and he has a problem with that?" I am now getting angry. This again. Shit! So irritating.

"No. Nor would we want you to change the fact that you are a woman," he says, and then adds, "look, I have two daughters, and if you are anything like they are – you would want to know the full story. Which is why I am telling you all this. And I believe that by now you have honed your skills of dealing with this kind of things… So – use them."

"Well, normally I just ignore idiots like that – I've learned that you cannot really change people. What else can a girl do? But you are telling me that I should not ignore this guy because of his position in the company. Surely, changing my behavior in deference to his position in the company is the essence of sexism. In a way you are asking me to give in to it. I am not going to go and try to charm him or anything. I mean, screw it!" I am angry and my usual reluctance to swear – especially at work and with my boss – slips. "If leading this project means that I have to schmooze someone like that – then I don't want it! Seems like a woman just cannot win."

"Calm down. Charming him probably would not work anyways," he mused. "I don't know Jasmine. I myself have never had this problem, so I cannot really say. Maybe the thing to do is to … well … maximize the profile of your other team members, use them to interact with Sam, or something? Get a feel for this Max guy – maybe you will not need much of Sam's help. I don't know…".

"You mean minimize my own profile?"

"Maybe. I don't know Jasmine. But I am sure that you will figure out some way – now that you have a full deck. You always do." And with that we walk back inside and go off to our offices.

I am glad that Dr Cz told me – I could hug the guy. But I do resent that I am getting this shit now – as if sheer delivery on this project is not enough. Back in Europe, in all those meetings with the EDA guys, I was the front, so to speak. I did the presenting and the talking, and I must say – I liked it. Now Dr Cz seems to be telling me that perhaps this is not the right way to play this particular game. Because of stinkin' Sam.

So, I put all this in a separate compartment in my head for now – to be revisited later – and focused on project delivery. First, I must make sure that we do a good job and that we have good results, and then I will worry about how to deal with Sam and how we roll it out. Cross that bridge when we get to it.

And that is the way it was for the next 3 months. It was a lot of work... Running all the different project coordination meetings, making sure that we are doing the analyses we set out to do, breaking down the tasks into bite size pieces and assigning them to suitable engineers, ensuring that these engineers had the time to do what they needed and when it needed to be done, getting the input information we need, driving the team to resolve any ambiguities, worrying that we are not forgetting something... And staying the course with all the conflicting pulls and pushes we were getting from what felt like everyone in the company... As they say, in many ways it was like herding cats – keeping everybody on task and in synch... A lot of work... And the thing that none of us fully anticipated – the sheer number of cross dependencies meant that we had many iterations of the analyses. We start with a candidate architectural "partition" – basically assigning specific content to each of the two dies in the SiP. Then we go and explore the implications of this partition on performance, cost and die sizes, power and temperature, pin out configurations, die and package routability, and even software stack configuration and compatibility with legacy products. And 'exploring' any of these dimensions often meant doing trial designs and many simulations. And doing the simulations, as always, required the right kind of constraints... And so on. All the experiences I had in working with the EDA guys helped – we had the tools, and I had something of a head start in their use. But these were not push-button exercises. And when we discover a no-go situation on any one of these vectors, then we go back and look at a different partition, or we tweak some other trade-off or constraint, repeat the analyses, and so forth... A lot of work... It like trying to unravel a giant ball of knotted string... A lot of work.

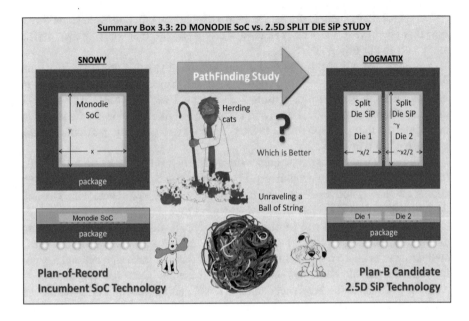

Through all this work, I learned an enormous amount. I felt like the Riddler from that "Batman Forever" movie – sucking up everybody's knowledge.

And it was not just the technical stuff but also some of the art of managing people and projects. Dr Cz helped me a lot here. He thought me that when you really need something done, you do not try to do it yourself, as dictated by that old parable, but look for the busiest person on the team and ask him – or her – to do it. Counterintuitively, but he is right: the most able and willing people on the team are usually the busy ones – for those very reasons. He also thought me that one of the most powerful tools for managing people is … pizza! He was being funny – but there was truth in what he said. If you offered an engineer $10 to stay late and finish a task, he – or she – would be offended. On the other hand, if you offer them pizza in appreciation of staying late and finishing a task, they happily spend half a night working. Go figure. The mind boggles as to what could be accomplished if corporate policies allowed beer on site. Of course, this is just a roundabout way of evoking team spirit and other such soft pressure points, rather than simple individual rewards. But, I prefer the way he puts it. For managing up – that is, managing your managers and their expectations – he thought me that, well, the way he put it, there are two kinds of people: Mrs. Yesif and Mrs. Nofor. The management always prefers a Mrs. Yesif. That is, the right response to any management request is "Yes, we can do what you ask, if…" and then list the conditions necessary to meet the request, and let them make the trade-off decisions. As opposed to responding with a "No what you ask cannot be done, for these reasons…," which precipitates an adversarial situation. Funny – but true. And for program management, he told me that there are only two rules. The first one is: no surprises. That is, the management perceives a project as well managed as long as issues are highlighted early – allowing them the option to address them. Failing to handle an issue on your own and then bringing it up when it is too late to do anything about it is the cardinal sin of project management. The second one is… the way he put it is 'not everything that is worth doing is worth doing well'. Again, funny – but right. You cannot do everything to perfection – there is no time for that. So, you have to select what to focus on and deal with everything else based on a cursory analysis, or judgment or – I hate to admit it – an educated guess.

I have also learned a lot about myself – something that I did not expect. For example, I am a Yesif – pretty much willing to try anything and focusing on how to do it, rather than thinking of reasons why not to. And Mariano is right – when I get involved with something, I go all-in and can get a bit obsessive. I guess that goes with being a Mrs. Yesif. I believe that this is why Dr Cz hired me in the first place. And, before this, I used to see myself as more of a hands-on individual contributor kind of engineer – happily hiding in my cube or in the lab. But, much to my surprise, I liked all the interactions with people. I shudder to think that I may be turning into a proper Valley-Girl; a giggly people person. But, these interactions were with really nice people – good engineers – and mostly meaningful and professional; maybe only slightly peppered with casual topics. That, and the fact that I genuinely like learning new stuff, actually makes me well suited for people leadership. The best way to motivate engineers is to be willing to listen and learn – and value – whatever they are working on.

Mariano has been great throughout these crazy 3 months – even helping out with home chores and stuff. He said that this is just his deposit in the 'Bank of Us' and that one of these days he will be the one who gets busy, and that then he will expect to make some withdrawals. Financial guy lingo, I guess. But I am liking this marriage gig. We also decided to postpone moving in together until his lease runs out – so, fortunately I did not have to deal with the moving hassles during this crunch.

And – with all that work and all that learning – we have found the sweet spot. We have identified a few configurations where 2.5D Split Die architecture made more sense than the traditional single die SoC! Lower-cost solution that met all the performance and other constraints. Taadaaa! I am so proud of the team and what we have accomplished.

So here we are, in the final Split Die PathFinding review meeting – presenting the results of all that work. Big room. Big meeting. Lots of people. Bill Mao and Alex, few other VPs, on down to everybody else. I decided to play it along the lines that Dr Cz suggested and am assuming a low profile. It is not just because of Sam and his stupid prejudices – it is because I really believe that the individuals on the team deserve it. They did the work, brought all their collective knowledge to the enterprise, and do deserve the limelight. So, I personally presented only one slide – an overview of the scope of the project that we all agreed on, back when we started – what feels like many more than the 3 months it actually was. And then I assumed a role of a scribe and let the engineers present their analyses, while I took notes. Of course, I orchestrated and choreographed it through the dry run meetings we had. And I am comfortable in the knowledge that Dr Cz and the team – the ones that really matter – are aware of my contributions. And we will let the data – the results of the analyses – speak for itself. The meeting went pretty much according to plan. Through the presentations we made sure that all our assumptions are identified, the underlying reasoning is brought out, and the results of the simulations are highlighted. And then we put up this final slide and made sure that it stayed up during the discussion:

	SNOWY	**DOGMATIX**
Performance	Plan-of-Record	✓ Meets Spec (slightly lower)
Power	Plan-of-Record	✓ Meets Spec (slightly higher)
Form Factor	Plan-of-Record	✓ Meets Spec (equal)
Cost	Plan-of-Record	⊃ Better (esp. in Yr1 and Yr2)
Risk (Sourcing)	Plan-of-Record	●✲Worse (it is new)

In keeping with the tradition, the plan-of-record version of the product with a single die SoC implementation was named "Snowy" – after the dog from the Belgian Tintin comics. Apparently, someone from Bill's team attended an IMEC workshop in Brussels, made a side trip to the Tintin museum – and picked that name for the next product. So, we named our 2.5D two-die version "Dogmatix" after the dog from the French Asterix comics. Another side benefits of my time in Europe – I now know their cartoons. Dogmatix surely trumps that puny mutt Snowy. And I loved the picture we found on the Internet – it conveys our net feeling that the 2.5D candidate is a no-brainer choice. After all, the whole purpose of the exercise was to address the growing cost of the single die SoC implementation. And we found it – a lower-cost split die solution!

Bill Mao then stands up and says "Thank you. An excellent job, team. And very very interesting results. You guys have clearly demonstrated that there is more to this 2.5D Split Die approach than we expected. It seems to be quite complicated. But you have managed to find a viable solution. Kudos and Thank You. The decision – whether to proceed with Snowy with the traditional single chip SoC implementation, or to embrace Dogmatix with a split die SiP solution – is not easy. Hard choice. So, I will take a position and am asking everyone here to argue against it…" He pauses a while, almost theatrically building up the tension. And then he continues "As much as it really pains me to walk away from potential savings that would roll straight down to our bottom line, I am going to continue with Snowy" … pause … "For two reasons. Firstly, I think that the risks with the split die technology are too high, and we cannot take any chances with our flagship product. Secondly, the benefits are transitory, and may be different with different yield learning rate, or with different wafer pricing... These factors are likely to change or can be negotiated away. Now, everyone, please tell me why I am wrong."

There is a palpable gasp in the room. Through the process of doing the analyses, most of us on the team became converts – we believed that the 2.5D SiP implementation was a way of the future and a better, more scalable solution, that would give us an advantage in the market. So, Bill's position was a big disappointment. All the enthusiasm was sucked out of the room in an instant – like a popped balloon. Stunned silence.

Being the lead, I felt an obligation to articulate the team's sentiment, so I pulled myself together and was about to stand up and argue the points, but Dr Cz reached out and held me down whispering "Wait. Keep cool."

Then Steve speaks up and says "Bill, I understand, and cannot argue against the two points you raised. But let me make a different argument. Point 1: we cannot fully de-risk a disruptive technology without a product. Point 2: if a product always follows the logic you identified, we will always embrace the incumbent technology solution, and never change our approach… Until some competitor beats us to the punch. And then it may be too late."

"True," responds Bill, "But I do not see such a competitor right now. Do you? We have a strong position in the market, and I think we can improve on it, while still pushing the existing paradigm. The problems we have with the cost structure trend is shared by all – so why should we be the ones to take the risk and lead changing the paradigm?"

And before Steve could react, Alex cuts in – always a consensus builder – and says "Is there a way of doing both? Sometimes, in the past, when we wanted to try out some new IP, or a technology feature, that we felt may be risky, we used to put it on a product die with an ability to fuse it out. That way we had a sort of insurance policy against it failing – paid for through extra area on the chip. On the other hand, if it worked, it was already baked into the product. Is there something like that, that we could do?"

"Hmm. I don't think so," opines Dr Cz, "implementing 2.5D integration is too disruptive a change. It is not just a matter of sawing a die into two to make a Split Die SiP solution, or choosing not to in order to have a single die SoC. We need to do an entire design – ground up – one way or the other, in order to reap the benefits of either approach. We end up with the worst of both worlds if we try to do both."

Silence.

"Well then," concludes Bill, "we will do Snowy. But please – do not let this decision discourage you. Do keep thinking and looking for alternative solutions. One of these days we may not have a choice – as you said Steve. Or, one of these days, we may have a product where the choice may be more attractive. But for this product, and for right now – for this year – we will pursue the incumbent solution. Thank you." And with that, he gets up and leaves, and the meeting is de facto over.

I am deflated, disappointed, and depressed, feeling like all that work was a waste. So, before everyone disperses, I stand up and announce "Team, hey wait. I think we did a really awesome job. I would like to thank you all… At Duffy's… I am sure that Dr Cz will be happy to buy us all a beer or two."

And then Steve pipes in "Guys guys... I agree with Jasmine. You all have done an excellent job... Great insights. The things we believed even as recently as 3 months ago now seem so… naïve… Fantastic learning – and, ultimately, that is what it is all about. Thank you, one and all. And, I totally agree and think that Cz here would be delighted to buy us all a beer."

"Well, who am I to argue against Jasmine and my boss. I guess I am buying," says Dr Cz, "6:00 at Duffy's."

So, we all – my team that did all the work for the PathFinding study, along with some of their management – gathered at Duffy's to share our disappointment, talk about the project, gossip about the stupidity of the management, and speculate about various political conspiracies. Regular engineer talk at Duffy's.

Sometime that evening Dr Cz pulls me aside for a chat – I guess he could see that I was bummed – and says: "You know, grasshopper, I think I know you by now. You are likely to go and beat yourself up because the proposal was not accepted – after all the work you have done. Let me tell you – this is just how these things go. Bill made his decision for Snowy, but that doesn't mean that the study had no value – for him, or for the rest of the company. If you think about it, most of us – you included – would probably make the same decision if we were in Bill's shoes. And it had nothing to do with Sam, or you being a woman, or how you conducted the project, or anything to do with you. So, relax and celebrate your accomplishment. You have executed well and learned all about project management. Good. You and your team have produced a huge amount of learning on this technology. Good. You have done a fine job of planting a seed for the future. Good. The dividends from this study *will* pay out sometime in the future – sometime, on some project. Trust me – that is how these things work… So, relax and celebrate." And with that he gives me an exaggerated pat on the back and we clink glasses. I like this Dr Cz. He is right, of course, and I will wonder and worry if there is anything that I could have done better that would have made a difference. We'll see.

Chapter 22
The Big Picture (Development of More-than-Moore Integration Technologies)

Today prof Dracula is visiting. We – ATI – have relationships with a number of prominent professors in the field. We typically sponsor research on specific selected topics and get first dibs, so to speak, on their grads. As a part of these, the professors often come to RoCo and deliver a lecture and talk about the status of the sponsored project. For this one I get to be the host – since I was a relatively recent student of his. Hard for me to realize that it has been almost 5 years since I graduated. Wow. At his request, ever since I graduated, I tried calling him by his name – Andrew – but somehow 'prof Dracula' seems to be hardwired in my head. Anyways, his lecture, as always, has been interesting and stimulating and was followed by a lot of Q&A. He is so good at that. I guess practice does make perfect. Afterward, the norm is that a few selected execs and the prof have a separate private lunch – an opportunity to get deeper insights from the prof. As a host, I get to participate in this one. Ha! Alex, Bill Mao, Steve, Dr Cz, prof Dracula, and li'l ol' me.

"You know, in the old days, these kinds of lunches were classy affairs. Some of the semiconductor companies had blue ribbon chefs on their payroll, and fancy dining rooms often adjoined some of the conference rooms. But nowadays – all we get is a catered lunch, water instead of wine, and plastic cutlery, in a regular conference room… Looks like Greek to me," says Dr Cz, informally chatting with the prof.. They are buddies, I believe.

"Greek is excellent," responds prof Dracula. "This is extreme luxury relative to my normal lunch at the U."

We all get our food and settle down, and prof Dracula squirms a bit and asks "Do you mind if I ask a non-technical question? It is for a new pet-project of mine… Stop me if I am venturing into improper territory."

"No, not at all… shoot," responds Steve, and then, jokingly, "well, depending on what you ask."

"Thanks. So, I believe that RoCo has been supporting your Advanced Technology Integration activity for a few years now, right? Would you mind sharing with me what are the principal big-picture barriers to product adoption, and the main challenges of the More-than-Moore technologies that you are experiencing?".

© Springer International Publishing AG, part of Springer Nature 2019
R. Radojcic, *Managing More-than-Moore Integration Technology Development*,
https://doi.org/10.1007/978-3-319-92701-5_22

He is looking mostly at Steve, so Steve takes a lead in responding, "Very topical question, professor. Alex, Bill – please correct me if I am wrong, but this is my take," and as is his habit, he gets up and writes on the white board.

1. *Axiom: Must Design for It*
2. *Iterative Intersection of Product with a sufficiently compelling Value Proposition*
3. *Transferring the Learning to Product Delivery Teams*

"First and foremost," he begins, "the key constraint that we have come to appreciate is that getting value out of these so called More-than-Moore integration technologies requires that you design a product specifically and explicitly for them – architecture and on down. They are not a kind of packaging technologies that you can select as an afterthought. In order to leverage the value proposition a product must be architected for it, ground up. This seems obvious to us now, but I do not think that we fully appreciated it when we started."

"Yes … MtM 101," says prof Dracula with a grin while writing in his notebook. He is a compulsive note taker, but his ability to retrieve an obscure factoid from some meeting that took place years ago is legendary.

Steve nods and goes on "For the second one, we are finding that standing up some of these Integration technologies requires a long runway. Developing the partnerships with the supply chain, designing the test vehicles, going through the learning cycles, and all that, takes years. By the time that we have something that is sufficiently baked to share with the product guys, the driver application that we had in mind has moved on. And getting the attention of the product guys for unbaked technologies is a losing proposition around here – they are all very busy. The learning generated along the way is obviously valuable, but typically by the time we are ready the technology does not meet the revised requirements of the application that was targeted at get-go. And, getting in front of that curve by anticipating future requirements, ends up with a concept driver-product that is so fuzzy and generic that it is hardly useful for zeroing-in on target technology specs. So, there is a conundrum there. We need a driver product to shape the technology, but that is a target that is moving faster than we can develop the process to desired level of maturity. So, we are finding that rather than developing technology solutions that converge directly with a product that we initially targeted, we really end up in a kind of an iterative loop. When we miss the initial application, we look around and find an alternative one, and then we scramble to tweak the technology for this new target, and so on. We are finding that we do this several times – each time learning more and getting a bit closer to a product intersection – until finally there is an actual convergence. It is like Brownian motion in a drift field – we are chasing a succession of target applications, each time tuning the technology a bit differently, until, finally, we have a technology solution that is sufficiently attractive for a product to swallow the risks and use it. So, in most cases, we just do not have that direct process-development-to-product-use convergence that is required for rapid deployment."

"Interesting," mumbles prof Dracula, jotting things down.

"Zeno's Paradox," says Dr Cz, but before he could expound on this, Steve says "Bill, Alex – what do you think?"

"Well, in defense of the product guys, there is no point spending a lot of BTUs on technologies that are too risky for use in real products," mumbles Bill, covering his mouth while chewing his food.

"True true. Nothing against product guys. You have a tough job. But the point I am making is that the development of these new and disruptive integration technologies is in reality turning out to be an iterative process. You first have to develop the technology up to some point based on an initial target application, and then you tweak it, and re-tweak it, for different candidate applications, until you finally converge with a right product." Steve explains and then elaborates, "we did not set out with this iterative convergence in mind. But this is how it is turning out. And the iterations are not necessarily distinct or consciously defined – they overlap and merge into a continuum of a development effort with moving target specs. In a way this sounds like a bad thing – like the 'creeping excellence' syndrome that all engineering managers are conditioned to avoid. But we are finding that this is exactly what we have to do in order to converge with a product."

"Interesting. So, now that you know this, would you do things differently?" asks prof Dracula.

"Good question. We haven't really thought much about that... I thought that having a corporate EVP act as an involved executive sponsor would help – someone high in the organization who would take an active interest and force a product design to converge more rapidly with process development. And to provide some shade, so to speak, during the period if and when we have to re-target the technology? ... Now I am not so sure. This chasing of a succession of potential applications means that we are continuously selling, so to speak – to the various internal product teams. This takes time and energy. Having an involved EVP may help to reduce that overhead.... Maybe we also need to invent some mechanism for storing the intermediate learning somehow. Right now, all the learning is pretty much in the heads of the engineering team – which is not the best way of feeding corporate memory. Of course, just documenting the learning and putting it on the shelf, also doesn't work – learning on the shelf just withers and dies if it is not used. I don't really know... Requires some thought."

"Very interesting. Thank you for sharing. And point 3?" asks prof Dracula.

"Ah yes," Steve responds, "that was unexpected too – but with all the brilliance of hindsight, it is obvious. We are finding that dealing with these More-than-Moore integration technologies is a multi-faceted challenge that involves analog and digital design, process, packaging, test, thermal, mechanical, electrical … everything really. That makes intrinsic sense, given the nature of the technologies. So, we grew ATI as a broad, multi-disciplinary, multi-physics team. But the flip side of that is that there is no suitable activity in the existing Product Delivery organization to receive the whole of our learning. It is much more insidious than just the usual gripe that designers typically cannot eat methodologies – they like tools. Product Delivery organizations are normally structured like silos of specialties. Which, by definition, means that no one silo can absorb the type of learning that we are generating.

Altogether, we are finding that handing off the learning to standard product delivery organizations is awkward, and we spend a lot of energy trying to map the learning into the silo structure... This also has some other unexpected implications. For example, we are finding that it is hard to get ATI people promoted with our peer-review mechanisms, since few of the peers get to see and appreciate the whole of their contribution. It is a sort of version of the old elephant-and-blind-men parable, and ATI people are just not getting the recognition that I feel they deserve."

He pauses, takes a drink, and continues "In addition, product guys need black and white pass/fail specs to sign-off and release a work-product; whether it is between the silos or to an external entity like the foundry. That makes sense, given the organizational structure and their job. But deriving the full value from integration technologies by definition calls for global optimization – something that a small multi-disciplinary ATI team can do, but that would be hard in a large rule-based organization. So, even if we managed to transfer our learning to some product delivery silo, I am not sure that they could run with it, so to speak, and apply the learning to some other MtM application in the future."

"Fascinating. Thank you. Same question, if you do not mind – would you do anything differently?" asks prof Dracula.

"Again, we haven't really thought much about that. Maybe we need to come up with some kind of a people rotation mechanism – where people along with the learning are cycled into the product delivery organizations? That is usually hard to do and is inefficient – but maybe we need something like that? Maybe we need to come up with cross-domain trade-off mechanism, so to speak. How do you trade off so many GHz of performance vs. so many degrees C temperature or so many square microns of chip area, or whatever? How do you trade off some value in, say, the package vs. some cost in say, Silicon or test? PathFinding simulations are a mechanism for estimating these values, but we also need a management policy that will define the rules for the tradeoffs across disciplines and physical domains. Each degree C is worth so many GHz, or something like that. In the absence of a management tradeoff policy, old metrics like performance typically trump new considerations like temperature. So, we end up doing all the thermal analyses just as a point of information, rather than to affect the design... There is a gap – and management needs to fix it somehow – in order to enable the silos to optimize and sign-off a design that leverages a More-than-Moore integration technology. Again, I don't know. I am just thinking aloud," says Steve, and adds "Alex? Bill? What do you guys think?"

"We should discuss this. Good points," opines Alex, giving a fairly generic no-commit comment. That seems to be his management style – building consensus rather than taking a stand.

"Well, thank you. This is truly fascinating. Would it surprise you if I told you that what you are experiencing is pretty much the norm across the industry? One of the advantages of being an academic is that the industry does not see you as a threat. So, I visit many companies and get to talk about all sorts of things. People are mostly willing to share with academics – like you just did. Thank you. I am finding that your experiences are not atypical at all. In fact, I think that what you are describing is a microcosm of what is happening in the industry. And I believe it is a problem that is intrinsic in More-than-Moore type of integration technologies," says prof Dracula, closing his notebook.

"Really? What do you mean?" Bill and Steve ask in unison.

"Well, think about it. In the good old days of Moore's Law, we had what you, Steve, call 'direct convergence' between process and design. Furthermore, we could do process and design development almost in parallel. Of course there were some handshakes between the two vectors, so to speak, and there was ITRS, and so forth – but, in principle the world of process technologies, including the foundries and their supply chain, could develop the technologies pretty much in parallel while the design world, including companies like yours and the EDA guys, etc., develop the leading-edge designs. That is because everybody pretty much knew what to expect – based on the history, the ITRS and other such things. As we progressed down the nodes we needed to sharpen our pencils and we established separate eco-systems targeting broad product families – like Logic vs. Memory. In the latter days of Moore's Law era we needed to get even more granular and Logic eco-system was further partitioned into Low Power vs. High Performance families, and so forth. But all along, we pretty much developed process technologies in parallel with developing designs, and we had that 'direct convergence'. Agreed?" says prof Dracula, getting up and walking up to the white board. Clearly assuming his professorial persona, he sketches something like this:

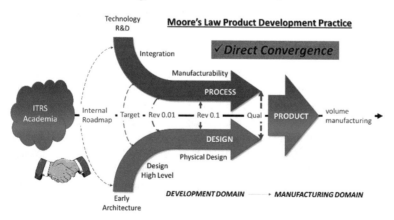

"You could represent the development time-line, something like this?" he says, pointing to his drawing. "Of course, there is a lot more going on, but at the high level of abstraction, the two development vectors converge with the qual of the lead product. There is obviously information handshaking going on before, keeping the two vectors synched up. Agreed?"

Everybody nods, and Steve points out "Big fabless companies actually have deep relations with the foundries...."

"Yes, I know," prof Dracula cuts him off, "so maybe the handshakes are more pronounced or more frequent, or whatever... But the two activities evolve pretty much in parallel and results in that direct convergence. At the level of abstraction I am thinking of, this chart applies to IDMs as much as to Fabless + Foundries," he explains.

"Now, let's think about development with More-than-Moore type of integration technologies," he continues, "there is no single mainstream process technology or a universal product requirement spec. There is no historical precedent to extrapolate. There is no industry roadmap like ITRS. There are no standards. There are multiple entirely different integration technology families that are thoroughly incompatible. Even the basic things that are analogous to wafer size in Silicon technology, such as the package substrate panel sizes, are not standardized. This makes sense only in an environment where process technology requirements are very product-specific. So, what do you use to target process development? It is like what you said, Steve – just like you feel that you must architect a product ground-up for a given MtM technology, the process development guys feel they need to develop a technology for a given specific MtM design. It's the nature of the beast at this stage of evolution. So, what do you do?" he asks and then elaborates, "two options...".

"Option-1 is to shoot blind and develop a generic process technology for some concept design. This may work, and if you are lucky your concept design will match some   real product needs. Then everybody is happy. If you are not lucky – which is much more probable – and your process technology does not meet any product requirements, then you can (a) wait until there is a product that does require the process that you happen to have, or (b) tweak the technology to meet some other requirement spec. It is another way of doing what you Steve called 'iterative convergence'. However, both of these propositions are difficult for foundries or OSATs. They are not getting paid while the technology is sitting on the shelf, or even worse, while they are investing incremental effort to re-engineer and re-qual it. So, this approach is an expensive hit-or-miss game – resulting in that iterative convergence mode," he concludes, looks around, and then continues.

"An alternative is to pursue Option-2, and to design first, i.e., develop a design to a level of maturity that is sufficient to drive process development. In that scenario the design sits on the shelf and waits for the process technology to be developed... I suppose that it is possible to do some mix of the two options. But, in principle, when the process technology is tied to a specific product, either the process waits for a design, or a design waits for a process. So, the parallel development and the direct convergence that we are used to seems to be very difficult to implement with More-than-Moore type of integration technologies." And he draws these sketches on the board.

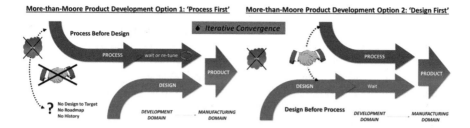

And adds "At least for now… Note that this is illustrating the point just looking at the two principal engineering vectors – process development and design. In reality for MtM integration there are so many more elements – things like packaging, test technologies, rf/analog components, sensors, passive elements, or what have you. How does one synchronize all those separate technology development efforts without a specific driver product? Something has to wait for other things to be developed. So, the two vectors are illustrative of the overall 'iterative convergence' point."

We all stare at the board, impressed by the simplicity of his drawing, and the global perspective, while trying to correlate our ground level internal experiences with his zoomed out generic view. Steve says "Wow, professor! This is excellent. Yes, this matches our internal experiences. Boy, I feel like a dummy. You make it seem so obvious, but we did not anticipate some of these hurdles."

"Well, hindsight is always 20–20," comments prof Dracula, "and, if it makes you feel any better, the entire industry is just learning this. This is a problem for the whole technology paradigm. Another symptom of the end of the traditional – and easy – Moore's Law paradigm. You guys are in fact somewhat ahead of the pack," responds the prof.

Bill clears his throat and asks "seems like this would have some implications on the corporate structure. If all the doom-sayers are correct and Moore's Law paradigm does come to an end, and we all have to do some version of More-than-Moore integration, than the distributed supply chain structure may be at a disadvantage relative to IDM? What do you think, professor?"

"Resurrect the old vertically integrated semiconductor company model?" adds Dr Cz.

"You mean because IDMs have both the design and the process under one roof it should make this convergence easier? Maybe. But in order to really have an advantage, an IDM would need to have all the key technologies under its single roof. That would require quite a span of expertise and infrastructure. So, all companies have a problem – but to varying degrees. Which brings up another issue – driven more by the maturing of the industry than by the nature of the MtM technology. In the old days, when a company needed to expand its technology portfolio, the fastest way was to go and acquire a smaller company with suitable expertise. Sort of analogous to Steve's idea of rotating people with right skills into silo organizations. Except that with maturing of the industry, the number of technology-centric small companies is drying up. It is harder nowadays to implement accelerated technology learning through acquisition, I think. The industry could form coalitions with the selected companies to bridge this span of technologies – along the lines of the kind of partnering that Steve, here, does. We have seen that sort of alignment arise in the past....," responds prof Dracula and then adds "As much as that would make the lawyers happy – all those contracts and NDAs and other legal paperwork – I think that there is a more fundamental challenge that needs to be thought-through."

He then gets up, erases the board, and again assumes his professorial role – seems a few inches taller to me – and says: "With the More-Moore paradigm the technology integration was, by definition, done at the Silicon level... And the 'technology integrator'...," he emphasizes with the air quotes, "i.e., the entity that produced a given Silicon technology – was the guy with a fab, regardless whether this is a foundry or an IDM. Right? I am now talking about the process technology vector itself, rather than the convergence of process with design that we just talked about. Anyways, to do this process technology integration, the Silicon Integrator has some degrees of freedom within the constraints that he has to live with. The way I think of it is like this," and he draws a chart on the board and continues:

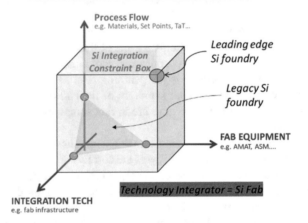

"So, the Silicon Integrator – in his quest to come up with the best Silicon technology – is constrained by the fab he has, the equipment he can get, and the integration flow he uses. So, he has to operate within a box defined by these constraints. I am purposely ignoring the human factor – because that is more subjective – and am therefore assuming that every Silicon integrator has equally talented team. Depending on the market that he is targeting – and on his purse – he goes and builds a suitable fab that includes all the integration infrastructure, the clean rooms, piping, shock absorbers, robots, filtering and what have you. He then selects the right equipment and uses the best process flow that he can come up with to create the best Silicon process technology that fits his constraints. Right? Basically, the integrator selects a point in the space defined by this constraint-box. So, a leading-edge foundry, for example, must have the best of everything and selects a point corresponding to the corner of this box. Best fab with best equipment and most advanced integrated flow, etc. And a fab-lite fab of some kind may operate on a more constrained plane, defined by his legacy fab, the equipment he has, etc., all within this box. And so on. Right? However you turn it, the guy with the fab – and of course with the silicon process expertise – is the integrator who produces an integrated Silicon technology that corresponds to some point within this constraint-box. Agreed?"

We all nod, a bit tentatively. Feels like he is going off into some bizarre academic left field?

He paces around as if thinking on his feet and says "Now let's transpose this concept to the More-than-Moore type of technology. The integration, by definition, is not done purely at the Silicon level, but involves multiple different technologies. It is not about integrating transistors on a piece of silicon but about integrating complete functions into some kind of a module. So, what are MtM Integrator's degrees of freedom and constraints? I would say his fundamental constraints would be (a) access to the packaging technologies, (b) access to the Silicon technologies – always, and (c) access to the Integration technologies – things that do not belong in either Packaging or Silicon domains – like the TSVs or uBumps or wafer bonding technologies. Agreed? So, his box of constraints would look like this," and he draws another chart on the board:

More-than-Moore Technology Integration Paradigm

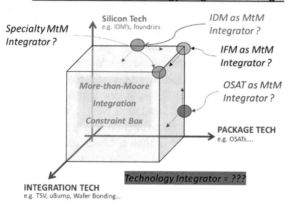

"So, the MtM integrator has to live within this box. The problem about iterative convergence with a product can be also viewed as a kind of a search for the sweet spot for a given application somewhere within this box. But the first question is who plays that role of that MtM Technology Integrator? What entity best fulfills the role that is analogous to the Silicon integrator from the MM paradigm? The industry is in the embryonic phases of the More-than-Moore era and no model seems to have emerged as a mainstream solution yet. It could be an IDM? Maybe. IDMs typically have excellent Silicon technologies, but a limited spectrum of packaging technologies. An IFM company like yours? You with your supply chain typically have access to a spectrum of Silicon and packaging technologies, but not necessarily the integration technologies. An OSAT? They obviously have the spectrum of packaging technologies but not much access to Silicon technologies. A Foundry? It is the flip side of the coin – access to Silicon technologies but not packaging. And so on." And he marks a spot on his box chart corresponding to each type of company as he talks about it.

"It is not clear to me that IDMs have an intrinsic advantage. Or for that matter, that any of the existing types of entities are a natural slam-dunk fit. In fact, it could very well be that there is room for a specialty entity that develops the best integration technologies – TSVs and micro-Bumps and such – and then offers MtM integration services in the open market. Like a foundry but for More-than-Moore technologies. There are small companies out there that are leaning that way today. A separate and dedicated 3rd party entity may have an advantage in that it could have the freest access to most Silicon and packaging technologies – not be encumbered by some competitive constraint. Foundries and OSATs may have problems in that space – getting hold of competitor's Silicon or packaging, respectively. IDMs and IFMs could fill the role of the Integrator, but would need to make an investment in the Integration Technologies. And such an investment may be more easily amortized by a 3rd party specialized entity than a captive provider servicing some internal needs. I don't know... Seems like there is a gap. At the current level of MtM technology complexity, this gap is filled mostly by OSATs, but I am not sure that they are a natural fit. I don't know…".

He pours himself some coffee, sits down, and concludes saying "I don't know… But I do believe that with the shift toward More-than-Moore type of technologies, the whole industry and the associated eco systems, will need to evolve new approaches to engineering development. A real paradigm shift. We are so used to Moore's Law that gave us such dramatic and predictable step-improvements every couple of years, enabling parallel development and a relatively direct process-design convergence. That may be hard to match. Maybe we need to get used to more incremental improvements vs. the step-function change every so often. A continuum of creeping improvements rather than distinct technology node stair-case… But, the two fundamental points are (1) it is not clear who will be the Technology Integrator for the MtM solutions, and (2) it is hard to develop technologies and designs in parallel. It is intrinsic and in the nature of the fact that these are cross-domain integration technologies, and that they are new – without much precedent. To enable the direct convergence, we would also need a whole family of new co-design methodologies, EDA tools, interfacing protocols – like Multi-Physics PDKs – and so on, to bridge the cross-disciplinary chasms. Big changes. And to enable the core technology development we may need to re-think the R&D model – maybe a revised R&D Consortium model. Or invent a different version of ITRS … It is new to all of us."

We all sit there a bit awed and amazed. Everything he says makes sense and is so big, somehow.

"Wow, Professor! My head hurts…," starts Dr Cz, "the implications of what you say are huge. You are suggesting that big shifts in the industrial landscape are in the cards."

"Yes. I believe so. If this More-than-Moore paradigm does take root, then it would make sense for the industry to realign to the new reality. Over time, of course. Possibly beyond my career span… I am not sure that my crystal ball is all that clear. I am just thinking aloud – and gathering the data. This is my current pet-project," responds prof Dracula.

"Excellent discussion, professor. Thank you" Alex cuts in "I must apologize, but I have to run." They say good-bye and Alex leaves. With this everybody looks at the clock, and Bill and Steve collect their stuff, shake hands with the prof, mumble something about great insights, and leave for their respective meetings.

Dr Cz and I then walk the prof back to the lobby, chatting about the various other companies pursuing similar efforts and some of the recent results that have been published. We thank him, of course, and he asks if we would think about coming to the university and sharing some of our learning with his class. Dr Cz volunteered me for that. And we say good-bye.

That was a great meeting, but I am quite disturbed by the implied messages. I feel like the discussion revealed a gloomy future for the technologies I was working on. So, I follow Dr Cz to his office and ask him "So Dr Cz, what do you think? Are we barking up a wrong tree? Have I just wasted 5 years – my entire career – working on a dead technology concept?" I ask.

"What?" he responds "Good grief – not at all. In fact, I think it is exactly the opposite. That was a good discussion and it brought into focus the challenges that More-than-Moore Integration technologies face. True. But if you zoom out, the way I look at it with my rose-colored-glasses, is that we know that Moore's Law is drawing to a close. We don't know when – but we know that that way of doing technology is drawing to an end. You have just invested 5 years in a technology paradigm that will likely succeed it. We don't know exactly in what form or when – but we know that a change is coming. So, you grasshopper, have a 5-year lead over your contemporaries who are working on the More-Moore technology paradigm. Probably more because you had a chance to look at many different aspects of a new paradigm at the ground floor. In my view…".

"You think? Well, hope that you are right," I say, feeling a bit better and encouraged. Although Dr Cz never fails to see a positive side to anything.

"Much more importantly, Jasmine, when are you off?" he asks.

I brighten up. "Well, the wedding is this Saturday. You are coming, aren't you? You promised…".

"Yes, of course. Would not miss it," he says, and I continue "But I am taking time off after tomorrow. A bride needs her beauty spa treatment and other such essentials, you know. And then we are off to The Big Easy for our honeymoon" grinning even more.

We started out wanting to keep the wedding fairly small and private – but this seems to be losing proposition. I guess that is the price we have to pay for coming from relatively large families. And then there are friends and colleagues which cannot be omitted. And pretty soon it is looking like a big wedding. It turns out that my cousins outnumber Mariano's, so we agreed to have the wedding here rather than in Peru. Nevertheless, the 'delegation' from his side of the family will be sizable. We jokingly refer to them as 'the Incan invasion force.' Hope it all goes well – no fights between the Incas and the Aztecs or the Persians. The wedding and the reception will be in Carlsbad – very nice. My mom and Mariano made most of the arrangements. She seems to be really into it, and Mariano has the patience for it – and for her. I am amazed how much there is to do, to organize, to think about, to negotiate, and what have you, and am happy that they have handled most of it. Seems to me that the whole affair is turning out to be almost as complicated as designing and building an IC.

That analogy makes me think. The last 5 years – more if I take in account my graduate years – have been all about work for me. I am feeling good about this. I really love what I do – never a dull moment. And I am feeling pretty secure in my career – prof Dracula's gloomy predictions notwithstanding. But – maybe this is a sign of aging or that proverbial biological clock – I am also feeling that there is more to life than work. Or, more accurately, I am feeling like I want to make sure that there will be more to my life than work. I forget who, but someone told me once that few people on their deathbed wish that they had spent more time working. I believe that. So, I am ready for those other things – with Mariano. Maybe some travel – see the world? Maybe build a nest – get a house and make a home? Pop out a kid or two? I guess I am all grown up now and know that there is a time and a place for it all. Not a choice of either-or but when. Yaaas!

Chapter 23
a Businessman: Mao's Musings (Circa + 5 Years)

I am a businessman. People think that I am an engineer – which I am by training and profession – but in fact I am a businessman. I am 'the money.' The thing that excites me is not making the greatest chip ever, or achieving the highest performance in a smallest die size, or any of the other metrics that turn on normal engineers. The thing that excites me is making money – or more specifically, making a product that makes a lot of money. It has always been that way for me. Ever since my father – back in Taiwan – shared his 'do not look forward, do not look backward, look where the money is' pearl of wisdom. He was referring to bikes and motorcycles, but the 'pearl' stuck with me. Life was hard for him. He and mother came to Taiwan in 1949 with the Chiang Kai-shek's migration, and things were difficult. He was not some general or a politician – just a man protecting his family from communists and war. So, after moving, he started a bicycle shop with the little money that he had saved up. With a lot of hard work, and maybe some luck, he expanded the business first into motorcycles and then cars. He ended up owning a successful chain of dealerships and provided a good living for the family; put both me and my sister through school – first National Taiwan University – and then graduate studies in the USA. But back then, when I was a child, he worked hard – 7 days a week, 12 to 16 hours a day. I remember his hands – always oily, with the black grease permanently stuck under his fingernails and in the cracks of his skin. And our home above the garage that was completely full of broken bicycles and parts. Everywhere – except in the kitchen. Mother would not allow him there. At one point, just as he was expanding the business into motorcycles, there was a flash fad for more powerful motorcycles – even models from Europe and the USA. James Dean rebel-without-a-cause look, sexy girls, and big bikes on billboards everywhere. Fashionable. At the time he had an option for an exclusive dealership of BMW motorcycles – seemed like a great and timely opportunity. But he decided to decline it, saying "do not look forward, do not look backward, look where the money is." He stayed with the bicycles and low-end

© Springer International Publishing AG, part of Springer Nature 2019
R. Radojcic, *Managing More-than-Moore Integration Technology Development*,
https://doi.org/10.1007/978-3-319-92701-5_23

Japanese motorcycles for another 10 years – and did very well. So that stuck with me. I worked hard in school and then studied engineering – because that is where the money was supposed to be. I came to USA, worked hard, and started my first job as a design engineer – because that is where the money was. I changed companies and assumed different positions – because that is where the money was. And now I manage a team and a product line like a businessman – following where the money is. Seems like his pearl of wisdom has served me well.

Engineers – I do respect them. Let's face it – without them the world would not be what it is. The talent, the inspiration, the drive, the vision, and always – always – the hard work. I admire and respect that. But engineers need to be managed. Sometimes I feel like they do need a resident adult to make the business work properly. Left alone, they will go and massage an idea, improve upon it, enhance it, and perfect it forever – and never make a product. They will always strive toward making a Rolls-Royce, except that most people want – or can afford – a VW. And no one can make a Rolls-Royce fit a VW budget. Normally engineers focus on the 'better product' side of the equation. But history is full of stories of companies with better technologies – and arguably better products – that failed. Like those poor guys at Mostek – one of the companies I worked for early on. History has it that the fortunes of the company, that at one point absolutely dominated the DRAM market, collapsed due to price competition from Japanese and Korean makers. That is true. But, the rest of the story is that we pursued a number of great, advanced, innovative, and excellent technologies before their time – and were therefore late to the market, with an uncompetitive cost structure. Our market share collapsed, and no one cared that we had a cooler product, with some really elegant advanced technologies. Or the classic Betamax vs. VHS and the early Mac vs. PC stories – now a subject of many business school case studies. 'Better' did not mean market success. Industrial landscape is littered with failed companies who had a better product. So, am I going to trust a business decision to an engineer? A starry-eyed engineer – like that Jasmine girl – all inspired and awed by some new technology? Like this 2.5D integration? Not likely! I need and want them to do what they do so well – invent and analyze and evaluate and assess and ideate and all that …. But I will make the business decision, and I will decide what goes into a product. Not because it is cool, or better, or even best – but because it will make money! That is what the company expects of me and what I do well. They need me – a businessman – to weigh the cool factor of technology vs. everything else.

On the other hand, there are an equal number of horror stories on the flip side of the argument – companies that failed, or lost market position, because they did *not* embrace a new technology sufficiently quickly. Starting from Fairchild – a company that arguably spawned the entire semiconductor industry – but did not hop on the bandwagon of polysilicon gate MOS technology quickly enough, and lost its leadership position, to Texas Instrument and Intel. Or Motorola – another iconic company, and a company I also used to work at – who ended up diverting too many resources to feed an incumbent bipolar technology – and lost position in the growing CMOS technology market. And so on – in our business with rapid technology evolution – you do not have the time to doddle. So, the technology trends *are* important – sometimes precipitating a live-or-die paradigm shift – and it is vital that we keep abreast of them. We cannot rest on our past – or even current – success. So, I do need engineers to do their thing. I do need an activity like Steve's ATI to look forward and to keep a finger on the pulse of the technology. And I do intend to listen closely to whatever they bring me. Sometimes, where the money is at today is different than where it will be tomorrow. Sometimes, moving from bicycles to motorcycles and then to cars makes sense. Every now and then, the technologies that the engineers bring me actually do change the world. Every now and then, there is a black swan among a flock of white ones – the one that really changes the game. They need me – a businessman – to spot the black swans, keep a cool head, and decide when a change should be embraced.

Then, there are also market trends to be considered. There are many examples of successful companies failing to perceive a shift in the market – and paying the price – sometimes the ultimate price of going out of business. Like DEC and SUN – both successful companies making excellent products and dominating their markets in microcomputers and workstations, respectively. But both of these companies were heavily vested in one market paradigm and failed to perceive – or accept – a shift in the market to a new paradigm – desktop PC, in their case. Or Eastman Kodak or Nokia – also classic business school case studies of once great companies who failed to embrace a shift in the market. It could happen to anybody, anytime. So, they need me – a businessman – to weigh the market factors and to decide on right investments at the right time.

And, of course, there is also the corporate dynamics to be considered – I will not call this politics. Investments, whether in terms of resources or money, in one product line must be balanced vs. investments in some other product line. Schedules and target revenues or margins must be viewed in context of the overall corporate performance – and how this plays on the Street. Viewed individually, this may, or may not, look like it makes sense – but viewed together it should play like a fine symphony delivered by a well-directed orchestra. That is, of course, assuming that the company is well managed and has a good strategic plan. I believe that RoCo does – which is why I am here, of course. And, navigating these dynamics requires dealing with people and personalities. That *is* politics – but a reality that has to be dealt with. Understanding these dynamics is a part of the constraints that must be factored in managing my business. I understand this. They need me – a businessman – to weigh these corporate factors and make the right trade-offs.

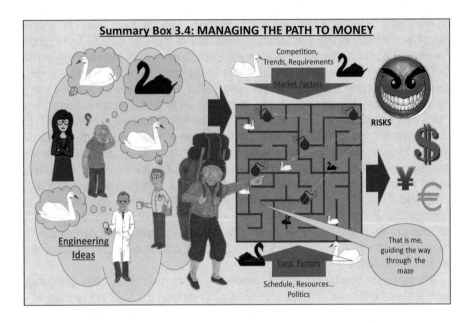

The one thing that most engineers forget – mostly based on wishful thinking – is *risk*. Risk in development, risk in integration, risk in manufacturing, risk in sourcing, or even risk of market acceptance. Engineers may wish it away or genuinely convince themselves that it is manageable – but I have been around long enough to know that it is real – and if you leave *anything* to chance, anything at all, it will get you. My product line competes in a market worth about $1B per year. So, am I going to roll a die and gamble anything more than I absolutely have to with a cool $1B on the table? No! They need me – a businessman – to weigh the risk factors and pick a right path between too good vs. not good enough, too soon vs. not soon enough, too expensive vs. not enough margin...

That is why they pay me the big bucks!

I run my product line as a businessman. I think of it as a separate company with its own profits and losses – rather than as a group within a larger organization. When I sat down and thought about my business risks, I realized that there is a potential problem looming out there – with the growing die cost and customers strong resistance to higher ASP's. So, I shared this with Steve. He and his ATI group are doing a good job – a necessary job – of looking out in the future. I am busy enough worrying about the current product and the next one that we are designing, and do need someone to warn me about things that may be looming around the corner – beyond my own horizon. That is his job. And when he suggested that we look at the 2.5D integration possibility to address my concern, I was all for it. There is a lot of buzz in the industry about the end of Moore's law and this 2.5D and 3D integration technologies, so it made a lot of sense to me to dig into it. I was a bit surprised when he picked Jasmine to be the lead. She seems young, and, in my opinion, Sam would have been a better choice. But it was his call. I value good relations with other groups much more than I care who is leading this kind of a study, or pleasing Sam. I don't know what Sam's problem was – and frankly do not care. It does not affect my business. And she, and the team, did a great job. I am really glad that we did the study. Interesting that the core idea of split-die architecture is so simple – two small die yield better than one big one. But the actual implementation of 2.5D technology is in fact quite complicated. I think that her study demonstrated nicely that adopting this approach would be a disruptive change. We would need to architect our product explicitly and specifically for 2.5D implementation and change many standard practices in design and test and yield management... This kind of change worries me. It is not like the usual scaling that we do – going from node n to n+1. That is hard enough as it is, but we have done that kind of technology hopping for years – and we know what to expect. This is different. I don't know what to expect. There could be many surprises. Frightening.

I know that there are companies and products that are embracing this 2.5D technology. But I also know the people at those companies who made their decision. They are like me – businessmen. They did not embrace the 2.5D technology because they want to have the bragging rights for being the lead adopters, or because they want to squeeze out an incremental bit of perfor-mance or power, or because they thought it was cool. They embraced the 2.5D technology because they felt they had no choice. There are some attri-butes of their products that just hit a wall when implemented in the traditional way. Maybe they just could not meet the core spec requirements – the basic target performance and power specs – using the single die SoC approach. Maybe their die size was getting to be so big that they spilled off a mask reti-cle, or they were getting no yield. Or something. But I do know that they took the risks of embracing 2.5D integration because they hit a wall and had no choice. That is how a big and frightening change like this usually happens.

For us, on the other hand, the split-die architecture benefits that the team identified were incremental – and just not compelling enough. If there was, say, a fixed 25% benefit – a 25% reduction in the cost structure that would roll down to my bottom line – then I would be tempted to consider it. But the 10%-ish that they predicted in year 1 is just not worth the risk. After all, the trend that I was concerned about hurts everybody equally – me *and* my competitors. And since I own the market that I am in, I think that we should not be the lead adopter of a disruptive technology. Why would I want to take a lead in changing the paradigm that we are so good at? In fact, sticking to the incumbent approach may be a good thing for us – the increase pressure on the margins may weed out some of my competitors. We are more likely to be able to withstand that pressure than they are... So, I think that we will turn the crank at least one more time and do what we have always done – stick to a single-die implementation, migrate to the latest CMOS technology as soon as possible, and then squeeze as much area out of the die as is possible. And for the rest, we shall just push our supply chain a bit more and see if we can negotiate a better deal.

Besides, there are the macro-trends to consider. So far everything – our internal analyses and the general market wisdom – points out that this 2.5D/3D integration is a pretty disruptive change. That prof even seemed to think that the entire development infrastructure will need to be realigned before the market can really adopt the new More-than-Moore paradigm. Wow! That is messing around with the very foundations of the industrial structure. Even if he is wrong – after all the industry rarely behaves the way that the academics expect – 2.5D/3D technology is certainly not something that can be easily dropped into our standard practices. Usually a technology disruption of this order of magnitude is propelled by some disruptive change in the market. Some new application, or new reality of some kind, is needed to drive a technical change. A new wave of entirely different applications – like the PC and the Internet or the smartphone a few decades ago … Typically… And when I look around, it seems to me that the mega trends in the market – of the magnitude that would be required to propel a disruptive technology – are IoT and AI.

IoT devices by definition integrate things like sensors, some processing power, and some communication capability into a single unit. This feels like a natural fit for the More-than-Moore technology paradigm. On the other hand, I suspect that IoT devices may be too cost sensitive to drive a disruptive technology change. Devices that aspire to be integrated into home appliances – toasters, fridges, maybe even groceries, and god knows what else – will have to be cheap. High volume – but very cheap. Usually the technology change is driven by high-end applications that then trickle down to the cost-sensitive applications. Usually… So, unless the industry invents a new way of funding IoT devices, it seems unlikely that they will lead the way to a new technological paradigm. Maybe some kind of a subsidy from the software end of the business – after all the real value is in the big data, not in the actual IoT devices. Maybe?

On the other end, Artificial Intelligence is needed – and is a result of this big data. AI feels like it will be a high-end application. The kind that has the margins to drive a technology paradigm shift. Maybe. In fact, the type of applications that are the early adopters of 2.5D integration are AI-ish… Like GPU processing? And GPU processing is a good fit for AI – for now. On the other hand, it could be that an entirely disruptive technology will be deployed to solve the AI challenges. Something like this quantum computing that people are talking about. The little that I understand of it, it feels like QBits might be a more natural fit to AI. GPU processing and the Artificial Neural Networks are a roundabout way of producing an answer that is a 'maybe' that seems to be needed for AI – rather than a simple 'yes' or a 'no' that is associated with conventional computing. QBits are intrinsically and naturally suited to a 'maybe.' So QBits may emerge as the ultimate AI solution – making the 2.5D implementation with GPU an orphan technology. Maybe…

Technical Background Box 3.2: IoT and AI

- *Internet of Things* (IoT): a network of physical devices equipped with suitable sensors, some local processing power (vs. processing power in the 'cloud'), and connectivity to the Internet, which are embedded in many modern appliances (e.g., thermostats, lights, speakers, toasters, ovens. etc.) and that will constitute the smart home, smart car, smart factory, smart cities, etc. of the future. IoT is expected to be the driver of the next big wave of proliferation of electronics – bigger than the PC and the Internet or the smartphone were in the past – and is expected to vastly outnumber the number of humans connected to the Internet. Next big thing!
- *Artificial Intelligence*: (AI) a 'new' methodology for extracting information from massive amount of data (as, for example, generated by IoT), based on a methodology and algorithms that can learn from (and make predictions on) data – as opposed to following strictly static and explicit program instructions. Voice and facial recognition, self-driving cars, etc. are instantiations of some form of AI. Also the next big thing!

Note: Material in the gray boxes is intended for those who are interested in more semiconductor technology and/or industry background information and may be skipped by those who are not.

But all that is not now – not this year or the next. So, in the meanwhile – until it all becomes a bit clearer – let's keep a real close eye on the technology and our competition. If one of them jumps, then we may have to follow. That is why we have Steve and the ATI team. To me, the value of the study that they did was to help me decide to *not* go the way of 2.5D integration. At least for now. That is important and valuable for my business. Sometimes not doing something is as important as deciding to do something. Like my father deciding to not accept BMW dealership before its time.

Good! They needed me – a businessman – to make these trade-offs. I am comfortable with the current decision. We are not looking forward, we are not looking backward, we are going where the money is.

Printed in the United States
By Bookmasters